COMBUSTION AND GASIFICATION OF COAL

A. Williams, M. Pourkashanian, and J. M. Jones

Department of Fuel and Energy
University of Leeds
Leeds, UK

N. Skorupska

National Power
Swindon, UK

Taylor & Francis
New York

USA	Publishing Office	Taylor & Francis 29 West 35th Street New York, NY 10001-2299 Tel: (212) 216-7800
	Distribution Center	Taylor & Francis 47 Runway Road, Suite G Levittown, PA 19057-4700 TEL: (215) 269-0400 FAX: (215) 269-0363
UK		Taylor & Francis 11 New Fetter Lane London EC4P 4EE Tel: 011 44 207 583 9855 Fax: 011 44 207 842 2298

COMBUSTION AND GASIFICATION OF COAL

1 2 3 4 5 6 7 8 9 0

A CIP catalog record for this book is available from the British Librar̶

Library of Congress Cataloging-in-Publication Data

Williams, A., 1935–
Combustion and gasification of coal / A. Williams, M. Pourkashanian and J. M. Jones and N. Skorupska.
   p.   cm. —(Applied energy technology series)
   Includes bibliographical references and index.
   ISBN 1—56032—549—6 (alk. paper)
   1.   Coal—Combustion.   2. Coal gasification.
   I. Pourkashanian, M., 1955– .   II. Jones, J. M., 1964– .
   III. Title.   IV. Series.
TP325.W69 1999
665.7'72—dc21   99-26691
                                                 CIP

Printed on acid-free, 250-year-life paper.
Manufactured in the United States of America.

6721136

CONTENTS

Introduction   ix

Chapter 1:   An Overview of the Energy Contribution of Coal   1

1.1   The origin of coal   1
1.2   Coal as a source of energy   2
1.3   The general nature of coal combustion and gasification   11
1.4   Advanced clean coal technologies   15
1.5   Influence on greenhouse gas emissions   19

Chapter 2:   Properties of Coal   21

2.1   Formation of coal   21
2.2   Characterization of coal   28
2.3   Coal characterization and classification   39
2.4   Coal chemical structure   52

Chapter 3:   Pollutant Formation and Methods of Control   55

3.1   Formation and control of particulate material   57
3.2   Formation of carbon monoxide   65
3.3   Pollutants originating from sulfur present in coal (SO_x)   67
3.4   Formation and control of oxides of nitrogen (NO_x)   73

Chapter 4:   Combustion Mechanism of Pulverized Coal   86

4.1   Role of pulverized fuel combustion   86
4.2   Devolatilization of coal particles   89
4.3   Combustion of volatiles   101
4.4   Char burn-out   101
4.5   Development of carbon burn-out models for high levels of carbon burn-out   110

Chapter 5:   Combustion Mechanism of Coal Particles in a Fixed, Moving, or Fluidized Bed   123

5.1   Fixed- and moving-bed combustion   123
5.2   Atmospheric fluidized-bed combustion   130
5.3   Pressurized fluidized-bed combustion (PFBC)   138
5.4   Pressurized circulating fluidized-bed combustion (PCFB)   140

Chapter 6:   Industrial Applications of Coal Combustion   142

6.1   Coal storage, preparation, and blending   143
6.2   Combustion on fixed or traveling beds   145
6.3   Conventional and advanced pulverized coal-fired boilers   147
6.4   Equipment for the combustion of pf in power plants: burners and furnaces   149
6.5   General features of combustion and furnace design   152
6.6   Fluidized-bed combustion   160
6.7   Scaling criteria for burners and furnaces   166
6.8   Computational fluid dynamics methods   170
6.9   Computation of emissions   178
6.10   Modeling of combustion plant   179
6.11   Power station and other boilers   182
6.12   Fluidized beds and stokers   185

Chapter 7:   Two-Component Coal Combustion   187

7.1   Co-firing of coal with biomass or waste   187
7.2   Co-firing with natural gas   192
7.3   Combustion of coal-water slurries   192
7.4   Formation of briquettes and smokeless fuels and their use   208

Chapter 8:   Coal Gasification Processes   211

8.1   Coal gasification   211
8.2   Basic gasification reactions   215
8.3   Gasification methods   219
8.4   Detailed reaction mechanisms and intrinsic kinetics   228
8.5   Process problems and environmental considerations   235

References 238

Appendixes 246

The appendixes consist of a compilation of tables and statistical and other useful data relating to the applications and particularly the engineering applications of coal.

Appendix 1 SO_2 and NO_x emissions conversion chart   246
Appendix 2   Values of equilibrium constants   247
Appendix 3   Combustion calculations for coal   248
Appendix 4   Calculation of the products of combustion allowing for dissociation   251
Appendix 5   Flame temperature calculations   253

Nomenclature 254

Conversion Factors 257

Index   259

INTRODUCTION

At the present time the coal industry faces considerable challenges. These principally arise from the need to control the emissions resulting from mining coal, its transportation and utilization. Some coal contains sulfur, nitrogen, mineral matter, and other environmentally unfriendly species; thus, coal utilization results in atmospheric, land, and waste pollution. Coal faces competition from natural gas, a fuel more environmentally benign but in shorter supply than coal. It also faces competition from nuclear power and renewable energy sources. If coal is to remain a major energy source, then a number of technical problems have to be solved. The object of this book is to outline these problems.

The approach taken by the book is slightly unconventional. The problems and possible solutions are dealt with in the first chapter—in outline. This is also true for the pollution problems that are considered in a general way in Chapter 3. The nature of coal is tackled in Chapter 2. Thereafter, the technical aspects of coal combustion are considered in the remaining chapters, namely, the combustion of pulverized coal and the combustion mechanism of coal in fixed, moving, and fluidized beds. Industrial coal combustion applications are then outlined, together with other combustion applications, including co-firing, coal-water slurries, and briquettes. Finally, gasification of coal, a possible major clean coal technology of the future, is discussed.

The book is suitable for a wide range of scientists and engineers in the field of coal utilization. It is also suitable for courses associated with combustion science or engineering. It covers the background of the subject and sufficient detail to bring the reader up to date in the current areas of interest and debate.

Acknowledgments

We wish to acknowledge the assistance of our colleagues and families and the word processing skills of Mrs. H. Strachan and Mrs. S. Ogden. We also thank Mr. J.U. Watts (US DoE), Mr. M. O'Connor (National Power), Dr. D. Merrick (MMD Ltd), Dr. A. Jones (PowerGen), and Dr. A. Heyes (DTI) for helpful information. We are also indebted to many people and organizations for permission to use photographs and diagrams, namely, Dr. A.R. Marshall (Mitsui Babcock), Star-CD, FLUENT, CFX, Dr. M.A. Serio, Poof. L.D. Smoot, Dr. M.A. Wojitwoicz (AFR), Prof. K.M. Thomas, and my former colloques Dr. C.T. Chamberlain, Dr. W.A. Gray, and the late Mr. J. Macrae.

A. Williams

M. Pourkashanian

J. M. Jones

N Skorupska

1

AN OVERVIEW OF THE ENERGY CONTRIBUTION OF COAL

1.1
The Origin of Coal

Coal is a naturally occurring hydrocarbon that consists of the fossilized remains of buried plant debris that have undergone progressive physical and chemical alteration, called coalification, in the course of geologic time. Coalification is the process of metamorphosis that takes place under conditions of raised temperature and pressure and results in the transformation of the original peat swamp through the progressive stages of brown coal (lignite), subbituminous coals, bituminous coals, to anthracites and meta-anthracites. The level that a coal has reached in this coalification series is termed its "rank."

Coal consists principally of carbon, hydrogen, oxygen, and small amounts of sulfur and nitrogen and is mainly in the form of polycondensed aromatic rings. Carbon atoms included in these rings account for 70–80% of the total carbon present in the coal, for example, bituminous coals. The aromatic nature of coal increases with rank. Coal is usually found in conjunction with mineral matter, and it is this content together with rank that determines its commercial suitability as a fuel.

1.2
Coal as a Source of Energy

There has been a considerable increase in energy consumption and energy production over many centuries; the growth over the last 140 years is shown in Figure 1.1. The world has a mixed energy economy using various energy sources as shown in Figure 1.2, although it is dominated by the combustion of fossil fuels as outlined in Table 1.1. Over the years, coal has played a very significant role, as shown by the data in Figure 1.3 for the last 140 years as well as in Figure 1.4, which shows in greater detail the situation for the last couple of decades. Clearly there is an interplay between the different energy sources that varies from country to country and that changes over the years and is controlled by

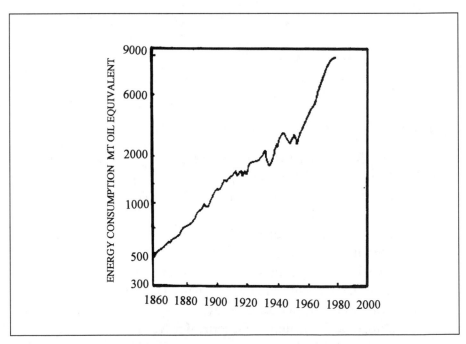

Figure 1.1   Growth in world energy consumption.

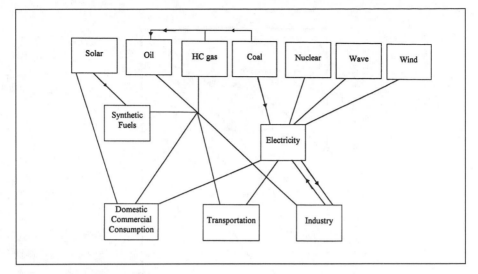

Figure 1.2 World's mixed energy economy.

TABLE 1.1
WORLD ANNUAL FUEL CONSUMPTION, 1997 (Oil equivalent)

		Consumption	Reserves	R/P
Oil*	3.4		124 Gt	41
Natural gas*	2.0	7.7 Gt	1.1×10^{12} m³	64
Coal*	2.3		579.4 Gt hard coal 444.3 Gt lignite	219
Nuclear energy [†]		0.6 Gt	20 Kt uranium	Large
Hydroelectric, solar, and other renewable		0.25 Gt	Infinite	Infinite
Biomass [‡]		1.7 Gt	> 450 Gt	Infinite (very large)

* *BP Statistical Review of World Energy*, 1997.

[†] Uranium Institute.

[‡] Estimated.

the availability of the different fuels. Availability, in turn, is determined by cost of supply, convenience of utilization, and environmental effects and costs. The main attraction of coal is that it is the world's most widely available fossil fuel energy source with a very large resource base and economically recoverable re-

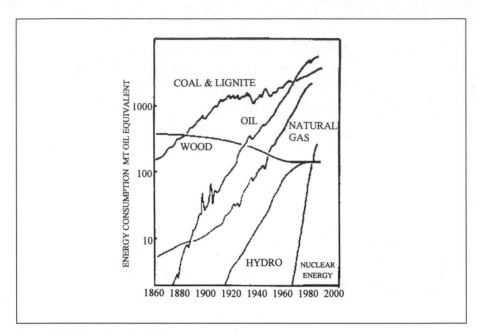

Figure 1.3　The changing pattern in energy usage.

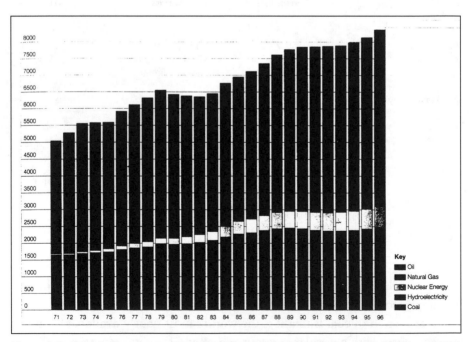

Figure 1.4　World primary energy consumption (*BP Statistical Review of World Energy*, 1997).

TABLE 1.2
DISTRIBUTION OF COAL RESERVES, 1997
(*BP Statistical Review of World Energy*, 1997)

	Anthracite and bituminous	Sub-bituminous and lignite	Total	Share of total	R/P
	Million tonnes				
US	106,495	134,063	240,558	23.3%	244
Canada	4,509	414	8,623	0.8%	110
Colombia	4,240	299	4,539	0.4%	141
Germany	24,000	43,300	67,300	6.5%	301
Poland	29,100	13,000	42,100	4.1%	209
Former Soviet Union	104,000	137,000	241,000	23.4%	*
United Kingdom	2,000	500	2,500	0.2%	52
South Africa	55,333	—	55,333	5.4%	255
Australia	45,340	45,600	90,940	8.8%	327
China	62,200	52,300	114,500	11.1	82
India	68,047	1,900	69,947	6.8%	212
Indonesia	962	31,101	32,063	3.1%	*
Total world	**519,358**	**512,252**	**1031,610**	**100.0%**	**219**
of which OECD	187,316	242,250	429,566	41.6%	230

* More than 500 years.

serves that are much greater than that of oil and gas. The world's proven coal reserves, unlike oil, gas, uranium, and many of the renewable energy sources, are spread relatively evenly throughout the regions of the world as shown in Table 1.2. It can be transported relatively easily. Consequently, coal has provided a relatively cheap source of energy.

In the mid-twentieth century, coal was the primary energy source in terms of energy consumption, and achieved this position as the result of exponential growth from the fifteenth century. One hundred years ago, wood was still the major source of fuel, being overtaken by coal only in the 1960s. Coal is a convenient fuel to extract, transport, and utilize because it is solid, relatively easy to break up or pulverize, and has a high energy density. The energy density is about 30 MJ/kg, which is significantly less than oil (40 MJ/kg) but vastly greater than other current forms of energy storage, especially electrical storage. The major problems with its use, especially in recent years, are the environmental problems and costs. First, the emission of smoke was a problem, which is now largely solved, then the acidic components SO_2 and NO_x manifest as acid rain were recognized in the 1970s as an increasing problem, again largely solved from a technological point of view. More recently the potential global warming effects of the greenhouse gases CO_2, CH_4, and N_2O can lead to difficulties, far from understood or solved at present. These result from the combustion of all fossil fuels but can be greater for coal. In addition, atmospheric pollution may result from the production, refining, and transportation of fossil fuels of all types, but coal causes special problems because of the release of methane due to mining.

At the present time the coal industry is of considerable size and is spread over 50 countries. The total coal production worldwide in 1996 was 3352.3 Mt hard coal and 1255.1 Mt lignite and brown coal, a total of 4607.4 Mt. The major producers of hard coal and lignite, respectively, in 1996 and listed in Table 1.2 were China, 1300, 50; former USSR 314, 92; India, 283, 25; S. Africa, 207, 0; the US, 562, 402; Poland, 136.3, 63; Germany 48, 187; Australia, 200, 50; and the UK, 50.5. Colombia and Indonesia are expected to become major producers shortly. Details of coal production are presented in Table 1.3.

Over the last decade, the volume of coal deposits classified as proven recoverable reserves and their availability has risen dramatically. At 519 Gt, hard coal (bituminous, including anthracite) proven reserves (1997) have advanced by very much more, and lignite (including subbituminous) deposits, at 512 Gt, have also increased over the values recorded 10 years ago. This growth is due primarily to the substantial rise in the reserves of the People's Republic of China. With some 610 Gt hard coal, China has probably by far the largest volume of hard coal reserves, although the proven reserves are 62.2 Gt. The US (106 Gt), the former Soviet Union (104 Gt), and India (68 Gt) account for more than 65% of total proven recoverable reserves. At present they produce approximately two-thirds of the world's hard coal. As far as lignite is concerned, some 60% of the proven recoverable reserves are concentrated in these three countries, although Australia and Germany contain substantial proven reserves of lignite.

Coal has thus further improved its long-term position as the world's most widely available fossil energy source compared with oil and gas (cf. Table 1.1). With current annual production of about 3.4 Gt of hard bituminous coal and 1.3 Gt lignite, sufficient reserves are available for several hundred years even if output should increase above the present level. Thus, because of the level of reserves, coal is expected to hold a supply saturation not far from its mid-1960s status as the world's major energy source in terms of consumption. The volume required to attain this position can be provided if production capacity is devel-

TABLE 1.3
DISTRIBUTION OF COAL PRODUCTION, 1997 (*BP Statistical Review of World Energy*, 1998)

	Million tonnes oil equivalent										Million tonnes coal	
	1990	1991	1992	1993	1994	1995	1996	1997	Change 1997 over 1996 (%)	1997 share of total (%)	Hard coal 1996	Lignite and brown coal 1996
US	561.4	539.9	539.8	505.5	551.7	549.5	566.2	573.9	2.3	25	562.2	402.0
Canada	38.2	40.0	35.5	37.7	39.5	41.0	41.8	43.3	3.6	1.9	40.0	35.8
Total North America	**603.2**	**583.3**	**578.5**	**546.7**	**595.9**	**595.2**	**613.1**	**628.1**	**2.5**	**27.1**	**610.7**	**437.8**
Colombia	14.8	15.3	17.0	15.7	16.2	18.8	20.8	23.3	12.2	1.0	28.8	—
Total S. & Cent. America	**20.8**	**21.8**	**22.7**	**22.0**	**23.5**	**25.8**	**27.1**†	**30.6**	**13.2**	**1.3**	**39.1**	—
Czech Republic	36.2	33.5	30.7	29.7	27.9	26.7	27.6	25.9	-6.4	1.1	17.6	63.9
France	7.8	7.4	6.9	6.3	5.5	5.1	5.1	4.2	-17.8	0.2	7.6	0.9
Germany	121.2	102.2	92.5	83.4	76.8	74.3	69.7	66.8	-4.3	0.9	47.9	187.2
Greece	7.1	7.2	7.6	7.7	7.8	7.9	8.2	8.2	0.5	0.3	—	58.4
Poland	94.2	90.6	85.4	85.0	86.6	87.8	88.3	88.1	-0.2	3.8	136.2	62.8
Spain	16.3	15.4	15.4	14.5	13.9	13.4	13.0	12.8	-1.0	0.5	13.7	14.2
Turkey	20.3	20.2	22.3	20.8	19.1	18.0	15.6	16.5	5.9	0.7	5.5	38.0
UK	56.5	57.3	51.4	41.5	29.8	32.3	30.5	29.5	-3.4	1.3	50.5	—
Other Europe	24.3	22.1	19.7	17.7	14.9	15.3	15.0	16.2	7.8	0.7	0.5	51.6
Total Europe	**401.3**	**371.5**	**348.6**	**322.4**	**298.6**	**297.2**	**290.4**	**283.7**	**-2.3**	**12.2**	**286.0**	**563.6**
Kazakhstan	67.7	66.9	65.3	57.3	53.4	42.6	39.2	37.1	-5.2	1.6	73.2	3.4
Russian Federation	176.2	154.8	148.4	134.8	120.7	118.1	115.0	110.0	-4.2	4.7	172.0	83.0
Ukraine	83.9	69.1	68.4	59.4	51.1	43.1	36.2	39.2	8.5	1.7	68.1	2.2
Total Former Soviet Union	**332.0**	**294.5**	**284.7**	**253.5**	**226.7**	**205.0**	**191.4**	**187.6**	**-1.9**	**8.1**	**313.5**	**91.6**
Total Middle East	**0.9**	**1.1**	**1.1**	**1.1**	**1.1**	**1.3**	**1.4**	**1.4**	**5.6**	**0.1**	**1.7**	—
South Africa	92.6	94.4	94.2	96.5	103.7	109.3	109.4	115.2	5.3	5.0	207.0	—
Total Africa	**97.5**	**99.3**	**99.2**	**101.2**	**108.5**	**114.2**	**114.0**	**119.8**	**5.2**	**5.2**	**214.7**	—
Australia	106.6	110.7	117.0	117.7	119.1	125.0	130.3	142.1	9.1	6.1	199.8	50.2
China	542.3	545.1	559.9	580.7	619.4	650.9	691.5	698.0	0.9	30.1	1300.0	50.0
India	103.3	110.9	117.3	121.5	124.6	132.6	143.1	151.8	6.1	6.5	283.0	25.0
Indonesia	6.5	8.7	14.2	17.0	19.1	25.5	31.4	33.5	6.9	1.4	48.8	—
Japan	5.5	5.3	5.1	4.8	4.6	4.2	4.3	2.9	-33.8	0.1	6.5	—
South Korea	9.1	8.0	6.4	5.0	3.9	3.0	2.7	2.4	-10.0	0.1	5.1	—
Total Asia Pacific	**815.2**	**829.7**	**859.7**	**885.6**	**927.6**	**977.6**	**1041.0**	**1069.5**	**2.7**	**46.0**	**1886.6**	**162.1**
TOTAL WORLD	**2270.9**	**2201.2**	**2194.5**	**2132.5**	**2182.0**	**2216.3**	**2278.4**	**2320.7**	**1.9**	**100.0**	**3352.3**	**1255.1**

* Commercial solid fuels only, i.e., bituminous coal and anthracite (hard coal), and lignite and brown (sub-bituminous) coal.
† Less than 0.05.

oped in good time. However, with lead times of 3–5 years for both open and near-surface extraction of geologically favorable deposits, any decisions have to be taken in good time.

However, a marked increase in prices can be expected in conjunction with this development, and this has to be viewed in the context of cheap world oil and gas prices. Today's low-price coal suppliers, with currently inflated capacities, are moving into more difficult geological conditions, so that new capacity must largely be generated either with difficult deposits or by creating a new infrastructure. Transportation distances are also generally rising between the producing countries and the users. Although cost forecasts are difficult to make, it is generally agreed that the cost increases incurred in exploiting new capacity will be significant. Independent of rising production costs, relative currency values in exporting and importing countries play a key role in determining prices on the world market, and these can be significant.

Pronounced technological advances in extraction and upgrading processes are unlikely in the foreseeable future, but the existing scope for introducing the highest possible technical techniques still remains to be fully exploited. New technological developments with proven commercial feasibility are essential in all sectors of the coal industry if this primary energy is to become sufficiently widely accepted to generate additional volume requirements. Environmentally friendly equipment for minimizing harmful particulate, sulfur dioxide, and nitrogen oxide emissions when burning coal represents state-of-the-art technology, but this has not yet been installed in most countries. Modern clean coal-combustion technology is also available for alleviating the CO_2 problem, by ensuring economical and high-efficiency energy use.

At present, approximately 46% of the world's electricity is generated by the combustion of fossil fuels. In some countries it is higher; for example, in India it accounts for 67% of the fuel mix. There is a consequential environmental impact due to the production of the fossil fuels, their transportation, and their combustion. At present, the solid fossil fuels are the major fuel used for electricity generation and are responsible for 58% of the electricity generated from fossil fuels, natural gas accounting for about 23% and fuel oils for 19%.

The energy scene, however, is apparently shifting toward a greater use of natural gas in many countries, and by the year 2000 natural gas could provide 25–30% of the world electricity output, while at the same time the amount of fuel oil burned will have decreased. The reasons for this are simply that, as already pointed out, although coal deposits are abundant throughout the world and are multisourced, coal's combustion by conventional means can cause a considerable amount of pollution. Thus, natural gas, which is a much cleaner fuel, is preferred by power generators and the chemical industry. However, the supply of the vast quantities of natural gas required do pose considerable engineering and marketing problems, especially in the long term. The expected decrease in heavy fuel usage is associated with market pressures to "crack" the heavy end of the barrel to supply gasoline to the rapidly growing transportation fuels market. As a consequence, heavy fuel oils in the future will be heavier (higher density, more viscous) and less abundant. The market is driven by the availability of the fuels, and this is reflected in their prices; in addition it is increasingly driven by environmental regulations that make clean fuels highly desirable. This latter factor is becoming increasingly important because it obviates the need to install a capital-intensive combustion chamber or flue gas cleanup equipment. Notwithstanding all these arguments, it is expected that coal will provide one-third of world electricity production by 2010 with a worldwide market over the next 15 years of about £500 billion ($800 billion) – which makes it a major industry.

The environmental impact of these fuels in relation to their utilization is briefly considered in the following section, together with an outline of the available methods of emission control. The pollutants produced are indicated in Figure 1.5, together with the impact on the environment.

Acidic gases The major atmospheric pollution issues involving combustion plant have been associated with the emissions of the acidic species nitrogen oxides (NO_x) and sulfur oxides (SO_x), and more recently, attention has been directed to the greenhouse-effect gases carbon dioxide, methane, and nitrous

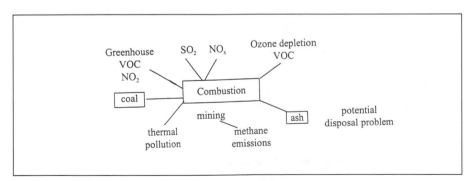

Figure 1.5 Pollutants resulting from coal combustion.

oxide. Emissions of SO_x and NO_x from fossil-fuel combustion are linked with a range of undesirable consequences, including acid rain, photochemical smog, and effects on health due to deterioration of ambient air quality. A 600-MWe unit generates approximately 20 Kt SO_2 a year for each 1% sulfur in the coal and about 10 Kt NO_x a year for each 1% fuel nitrogen in the coal.

Other pollutants   Other partially incomplete combustion products (termed PIC) are also present in low concentrations, notably carbon monoxide (CO), and unburned hydrocarbons (UHC or more commonly called volatile organic compounds VOC) are also implicated in the above effects, especially via ozone (O_3), which is formed as a product of reactions involving NO_x, CO, and UHC under the influence of sunlight. Toxic organic micropollutants (TOC) are also formed and these include PAH, dioxins, and furans and again result from incomplete combustion and the interaction with chlorine compounds. Aerosol particles are formed, which include soot, soot precursors, and airborne ash aerosols.

In addition there is the formation of the major part of the residual ash as a solid waste. A typical 660-MWe unit can consume about 6 Mt coal/year and can generate around 200,000 tonnes of ash a year (assuming 10% mineral matter in the coal). This is largely disposed of by landfill methods or use in the construction industry. However, some ash escapes with the flue gases as fine, or ultrafine, particulates that add to air pollution. Generally ash is considered as inert (although unpleasant), but it does contain trace elements that can be considered environmentally unfriendly.

Greenhouse gases   The greenhouse effect, whereby radiation of heat from the sun is trapped within the earth's atmosphere, has long been known, but only recently has it become the subject of intense international debate and policymaking. Although most major greenhouse-gas species, with the exception of chlorinated fluorocarbons (CFCs), occur in nature, fossil-fuel utilization is a major source of these pollutants. Carbon dioxide is undoubtedly the major greenhouse gas, but methane and nitrous oxides can also make a significant contribution, the latter also through its involvement in ozone formation.

New/newer clean coal combustion technologies will also be discussed in terms of the robustness of the technology, the fuel supply, and market situation. Estimates of emission factors for whole or partial processes will be given for these technologies, with particular attention being directed to cleaner coal combustion.

Finally, the amounts of atmospheric pollutants generated from energy production will be compared with natural levels and estimates given of the significance of their environmental impact.

1.3
The General Nature of Coal Combustion and Gasification

Coal utilization covers a wide area of applications, but in terms of classes it involves domestic and commercial heating, industrial applications, and power station or utility applications. Only the most basic coal-burning plant can deal with a range of coal sizes, and coal is subject to sizing operations. The method of utilization is related to the physical size of the coal, and this physical size is often referred to as "grade." The sizes of coals for particular applications are given in Table 1.4. The ash content from the different grades varies quite significantly. Often the coal grades are given trade names such as cobbles, trebles, etc.

Coal can also be gasified and the coal gases produced used for power generation or chemical synthesis, although at the present time this process is not extensively used. Coal gasification may be undertaken in fixed (or slowly moving beds), in fluidized beds, or in entrained gas flows using the same basic systems outlined in Table 1.4, although the fuel to oxidant ratio is changed so that only partial oxidation takes place.

Historically the method most widely used for combustion was a fixed bed of coal with the air flowing through a grate by natural convection and effectively a

TABLE 1.4
TYPES OF COAL COMBUSTION

Type of combustor	Type of air flow	Size (diam)
Fixed grate	Natural convection of air	5–15 cm
Moving (chain)-grate industrial stokers	Forced convection	0.56–1 cm
Atmospheric or pressurized fluidized bed	Fluidizing flow	1 mm^2
Pulverized fuel combustor	Forced-air flow	0.05–0.15 mm
Micronized coal combustor	Forced-air flow	~ 1 μm

poor control of fuel-air mixture, which also varies as coal is added to the grate. While this level of control can be improved by enclosing the combustion in a combustion chamber, there is still poor control of local fuel-air mixture because of the unevenness of the coal-particle size and bed depth. Some improvement can be achieved by automatic feeding and the use of moving grates. However, in the 1930s there was a move, especially in the US, toward using pulverized coal combustion. The advantage in using small particle size is that higher rates of heat and mass transfer can be achieved. A 100-mm coal particle has a surface area that is 1000 times greater than a 1-cm particle, and with consequent greater heat and mass transfer rates.

As a consequence, coal utilization, especially in electric power generation in Europe and North America, is now dominated by large pulverized coal-fired boilers, although conventional steam cycles are still in use. Generally the coal is burned as a pulverized coal powder (pf or psf) with particle sizes of about 50–150 mm together with air, although it can be burned as a pf coal (70%)/water (30%) slurry. The chemical composition of this bituminous coal is typically carbon 80%, hydrogen 10%, sulfur 1–3%, nitrogen 0.5–1.8% by weight on an ash-free basis, together 10–30% ash (inorganic components) with a calorific value of about 40 MJ/kg. The lignite or brown coal has a composition that is dominated by high water content, low carbon content, high ash content, and consequently, a low calorific value. Typical compositions of some brown coals are given in Figure 1.6; bituminous coals would tend toward the origin of this plot. Bituminous coal also contains approximately 20–40% "volatile" material, whereas brown coals contain much less.

Bituminous coal is therefore the preferred fuel because it is indigenous to many countries; see Table 1.1. It is generally a more satisfactory fuel than brown coal for a variety of reasons, but the principal reasons are that it has twice the calorific value of brown coal, does not contain much moisture, and burns more rapidly. Many countries, e.g., Germany and Australia, utilize both bituminous and brown coals.

The combustion processes that occur involve the following stages:

1. The coal particle enters the hot combustion chamber and heats up, resulting in the pyrolytic decomposition of the organic structure within the coal.

2. This pyrolysis process results in rapid devolatilization, giving a carbonaceous char particle together with gaseous tars and a combustible gas consisting of hydrogen and methane. The tars and the gases ejected by the

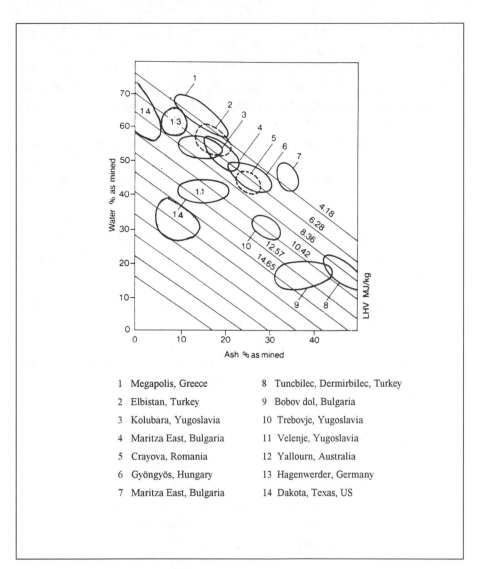

Figure 1.6 Variations between lignite/brown coal (after Moumdjian et al., 1984; IEA, 1978).

1 Megapolis, Greece	8 Tuncbilec, Dermirbilec, Turkey
2 Elbistan, Turkey	9 Bobov dol, Bulgaria
3 Kolubara, Yugoslavia	10 Trebovje, Yugoslavia
4 Maritza East, Bulgaria	11 Velenje, Yugoslavia
5 Crayova, Romania	12 Yallourn, Australia
6 Gyöngyös, Hungary	13 Hagenwerder, Germany
7 Maritza East, Bulgaria	14 Dakota, Texas, US

coal particles burn rapidly in the early part of the coal flame, while the carbonaceous char burns out more slowly leaving an ash residue. The combustion conditions, particularly temperature and residence time, determine the extent of remaining unburned carbon, which in turn determines combustion efficiency. The flue gas contains mainly CO_2, H_2O, and the residual nitrogen from the air.

NO_x and SO_x are minor components in the flue gases, typically about 300–1000 ppm of NO_x and about 500 ppm of SO_2 per %SO_2 in the coal. (A conversion chart is given in Appendix 1.) The SO_2 is formed simply by the oxidation of the sulfur present in the coal, present as both organic and inorganic compounds. The latter, present in pyrites, can be reduced before combustion by coal-washing or other forms of coal benefication. Unfortunately in many countries, such as the UK, coal contains sulfur mainly as the organic-coal compounds, and it is very difficult to remove these compounds by pretreatment. The SO_2 in the flue gases is simply formed by the oxidation of these compounds, although a small amount of sulfate is left in the ash.

The SO_2 can therefore be removed only by in-combustion-chamber reaction with limestone ($CaCO_3$), which is generally applicable only to lignites, or by flue gas desulfurization (FGD). The formation of NO_x is much more complex and involves:

1. A thermal NO_x route involving the reaction of oxygen atoms (O) present in the flame gases with molecular nitrogen

2. A fuel-N route where these compounds present in coal are pyrolyzed to give HCN, which in turn is converted to either nitric oxide or nitrogen depending on the fuel/air ratio. Fuel-lean flames tend to give NO, fuel-rich tend to give N_2.

In the last few years it has become possible to reduce NO_x emissions by about 50% by using two-stage burners. The first stage, just after the burner mouth, produces a fuel-rich mixture that reduces both the flame temperature and the oxygen atom level, while fuel nitrogen compounds are reduced to N_2; consequently, the NO is reduced. However, some NO is produced during the second-stage burn-out of the char, and thus two-stage burners are limited to only a 50% reduction in NO.

Fluidized-bed combustion provides an option. It was largely initially developed in the UK in the 1960s to the 1980s for both small-scale and large-scale power generation and is widely used. It offers the possibility of low NO_x formation coupled with in-combustion chamber SO_2 capture.

The limitations of the control methods available for sulfur- and nitrogen-based pollutants have led to the development of clean coal (or cleaner coal) technologies. These are outlined in the following section.

1.4
Advanced Clean Coal Technologies

In the last 25 years, several advanced power plant and solid fuel firing concepts have been studied in respect to their application as so-called clean coal technologies. This was initially prompted by the OPEC oil price increases, but increasingly in recent years by commercial pressures resulting from the competition from natural gas, and of course, these are enhanced by the environmental needs. Special emphasis has been placed on certain technologies that are expected to be capable of meeting the more severe requirements in terms of emission control and efficiency. In particular the following cycles are thought to be suitable to fulfill these criteria and are expected to be available for industrial-scale demonstration in the reasonably foreseeable future:

1. Advanced pulverized coal-fired boilers, and including pressurised pf combustion
2. Atmospheric fluidized-bed combustion (AFBC) and pressurized fluidized-bed combustion (PFBC)
3. Integrated gasification combined-cycle (IGCC)systems
4. Advanced integrated gasification, including fuel cell systems
5. Magnetohydrodynamic (MHD) electricity generation

Because of the current need to increase efficiency per se or to reduce CO_2 emissions, and to reduce emissions generally, clean coal technologies are receiving major attention at present. For this reason they are further outlined below.

Pulverized Coal Combustion

Current conventional coal-fired plants that use the Carnot steam cycle and are equipped with emissions control technologies for particulates SO_x and NO_x are proved and effective and contribute substantially to the reduction of pollutants from existing fossil fuel units. The next generation of coal utilization technologies has to improve economic and environmental performance by attaining much higher levels of energy efficiency and by using better methods of pollution control.

However, while much attention has been directed to clean (or cleaner) coal technologies that use combined cycle technology, it should be noted that advanced pulverized coal plants have been built at capital costs reported to be less than half of that currently envisaged for IGCC and for PFBC. These technologies, which are robust and flexible in operation, offer improved environmental performance. They are a baseline against which any other new clean coal technology will be judged.

It should also be noted that ultra-supercritical pulverized coal plants with the potential of electrical efficiencies around 45% are under development (Scott and Carpenter, 1996). For example, in Japan the Wakamatsu 50-MW demonstration plant has demonstrated two types of supercritical conditions, namely, 315 bars, e.g., 593°C double reheat, and 350 bars, e.g., 650°/593°C with double reheat. The strategy is to progressively commercialize the technology step by step at the following power plants: Kawagoe gas-fired plant (310 bars, 566°C double reheat), Hekinan coal-fired plant (240 bars, 538°/593°C), and Matsuura (240 bars, 593°/593°C with variable pressure operation capability).

Several plants in Germany, of about 500 MW, operate at 43–44% efficiency. The planned 700-MW Hessler project claims an efficiency of 45%, as does the 400-MW ELSAM project in Esjberg.

It is clear that the developments mentioned in this section are leading toward simpler, larger, more efficient, and cost-effective coal power plants, and provide the competition for CCGT applications with natural gas. Experience gained from the construction and operation of such commercial plants is essential to convince power utilities that the technical and economic performance of clean coal combined cycle technologies is comparable to existing technology. Operational performance parameters that are important include fuel flexibility, electrical load following capability, and RAM (reliability, availability, and ease of maintenance).

Pressurized Pulverized Coal Combustion

The pressurized pulverized coal combustion (PPCC) technology refers to the directly coal-fired gas turbine principle. Pulverized coal is burned at elevated pressures (> 20 bars) providing a high-temperature flue gas above the ash melting point. Depending on the feedstock, the flue gas temperature level may well exceed 1400°C. "High-temperature" gas cleanup systems are located downstream of the pressurized combustor in order to capture ash particles and volatile alkali species prior to the gas turbine. There are considerable problems imposed on the gas turbine, namely, erosion, corrosion, and deposition (Stringer and Meadowcroft, 1990) and ruggedized machines have been developed to cope with the problem. Flue gas desulfurization and deNO$_x$ processes are needed downstream of the boiler to meet the appropriate environmental standards.

This type of PPCC may achieve a net efficiency of some 50%. However, the efficiency could be raised because of developments in future gas turbine efficiencies.

In addition to the directly fired gas turbine principle, several indirectly fired cycles (IFC) are presently being investigated. These reach an advanced high-temperature ceramic heat exchanger to transfer the heat from the combustion section to a pressurized air stream that is the working fluid of a gas turbine. Thus, the gas turbine is not directly exposed to corrosive and abrasive combustion products. The ceramic heat exchanger tube will heat clean filtered air from the gas turbine compressor to high gas turbine inlet temperatures. PPCC technology still presents many problems and requires research and development in detailed understanding of various mechanisms related to pressurized combustion (e.g., chemistry, particle behavior, mass and heat transfer), retention of vapor-phase alkali species, and removal of fly ash, which may be molten. Material wastage (e.g., erosion and corrosion) of components exposed to the high-temperature corrosive environment and combustion, slagging, and corrosion behavior of various types of feedstock are also important. The development of robust high-temperature ceramic heat exchangers is also necessary.

Fluidized-Bed Systems

Fluidized-bed combustion of coal started in the early 1980s, but was largely used to raise steam or hot water. Essentially the aim was to replace conventional coal-fired boilers ranging from quite small units to power stations. However, it was realized that this technique, if used at elevated pressures, could be used with gas turbines. The advent of large industrial gas turbines made this increasingly possible during the late 1980s and into the 1990s.

Research programs in a number of countries then demonstrated that advanced clean coal technologies such as the pressurized fluidized-bed combined cycle (and the integrated coal gasification combined cycle) are capable of higher efficiencies, about 45% (based on a net electricity generated, LHV basis), and an enhanced environmental performance. They are still, however, currently in their development stage since the plants are relatively small (~250 MW), and unit capital costs are very high compared with conventional technology.

Commercial operation of a pressurized fluidized-bed combustion/combined cycle (PFBC/CC) plant was undertaken in Sweden to produce 135-MW electricity and 225-MWth for district heating since 1991, as well as demonstration plants in Escatron, Spain (80 MW, since 1990) and Tidd, US (70 MW, since 1990). At present, six pressurized fluidized-bed plants totaling about 1 GW are in use, and they could be widely commercialized in the next 10 years if they could meet the requirements of RAM and cost. Generally, the present plants are too complicated, and considerable simplification and cost cutting measures are required. At present the preferred types of plant have not been determined.

Coal-based Combined Cycle Technologies

A further development was the so-called integrated gasification combined cycle, (IGCC). IGCC technical developments are consequently focusing on higher efficiency, improvements in operation, maintenance, and reliability, and the lowering of capital costs. These include hot gas cleanup, cheaper air separation procedures, and more efficient combined cycle systems. Moreover, as awareness of the commercial opportunity grows, gas turbine manufacturers are increasing their marketing activities that target the early introduction of clean coal combined cycle technologies.

Studies on a plant in Finland by Ahlstrom Pyropower have led to a utility size demonstration of pressurized circulating fluidized-bed combined cycle (PCFB/CC) technology at the Des Moines Energy Center, US. These incorporate advanced cycle developments where coal is partially gasified in an air-blown spouted pressurized fluidized-bed (PFB), and residual char formed is burned in a circulating fluidized-bed combustor (CFBC). The gas fuels a combustor turbine, and heat from the char combustor joins heat from the turbine exhaust to raise steam for a steam turbine. British Coal has also developed a similar system, the airblown gasification cycle that, so far, has not been commercialized.

Japan has been developing an air-blown fluidized-bed and entrained-bed IGCC. However, after difficulties with both types of IGCC, Japan decided to develop the PFBC/CC and commercialize it first (Sugawara et al., 1998).

Oxygen-based Integrated Gasification Combined Cycle (IGCC) Technology

The processes previously discussed were based on the use of air. The chemical and gas industry has developed a number of gasification techniques for chemicals or gas production that can be applied for power generation. Several IGCC plants are now in various stages of commercial development worldwide. A brief overview of the status of coal gasification technology projects is given next.

The 250-MW Buggenum plant in the Netherlands was built by DEMKOLEC, a subsidiary of the power utility SEP. The plant uses Shell gasification technology, KWU/Siemens combined cycle technology, and an Air Products air separation unit. The net efficiency will be 43% and it is expected to be the cleanest coal power station ever constructed. The plant has been assembled and commercial operation started in November 1993. The plant is not subsidized and the additional power generation costs will be added to the electricity ratebase.

A 320-MW IGCC plant has been built at Puertollano in Spain that uses PRENFLO gasification scheme and started on natural gas in 1997. RWE in Germany has announced the KOBRA project at Goldenbergwerk, which will use the High Temperature Winkler technology. The size will be 308 MW and the unit will operate with lignite as the fuel. Several plants have been announced in the US, including Wabash River Repowering (265 MW), Polk County (265 MW), Delaware (250 MW), and Camden Energy (480 MW).

This means that all the key gasification types will have their technology commercially operational with power utilities before about 2010.

In addition to electrical power generation, gasification is also used for household gas supplies. Thus, in China a million households are supplied with a gas high in CO/H_2 by means of a Lurgi gasifier.

1.5
Influence on Greenhouse Gas Emissions

The utilization of coal involves a number of greenhouse gas emissions, namely, methane from the mining and grinding operations and CO_2, CH_4, and N_2O from its combustion. It can be approximately estimated that deep mining of bituminous coal results in the emissions of 15 m^3/tonne coal and surface mining results in 0.5 m^3/tonne coal [Smith (1997), Williams and Mitchell (1994)]. The major emission of a greenhouse gas—and a measure of its efficiency—is that of CO_2.

Considerable reduction in CO_2 emissions can be achieved by the use of these technologies. These are shown in Table 1.5, which is based on IEA Coal Research estimates (Smith et al., 1994).

TABLE 1.5
POSSIBLE REDUCTIONS IN CO$_2$ EMISSIONS FROM COAL-FIRED PLANTS (IEA)

Technology		Net plant efficiency (%) LHV	CO$_2$ emission factor (gC/kWh)			CO$_2$ reduction (%)
			Coal	Limestone	Total plant	
Conventional systems	Pulverized coal					
	Reference plant	36	252	6	258	0
	Subcritical steam	39	253	5	238	8
	Supercritical steam	42–45	202–216	4–5	206–221	14–20
	Ultra-supercritical steam	47	193	4	197	24
	AFBC					
	Subcritical steam	39	233	10	243	6
Combined cycles	IGCC					
	Demonstrated systems	38–43	211–239	0	211–239	7–18
	Advanced systems	45–47	193–202	0	193–202	22–25
	PFBC					
	Subcritical steam	44	206	4	210	19
	Supercritical steam	46	197	4	201	22
	Hybrid*					
	Subcritical steam	47–49	185–193	4	189–197	24–27
	Supercritical steam	52	174	3	177	31
	MHD					
	Subcritical steam	42–47†	193–216	0	193–216	16–25
	Fuel cells					
	Subcritical steam	47–60†	151–193	0	151–193	25–42
Combined heat and power	Pulverized coal					
	Subcritical steam	85†	107	2	109	58
	PFBC					
	Supercritical steam	91–92	99–100	2	101–102	61
	Subcritical steam	86	106	2	108	58
Fuel blending	Pulverized coal					
	Subcritical steam, coal/natural gas (85/15)	37	231	5	236	9
	Subcritical steam, coal/oil (47/53)	36	224	5	229	11
	Topping GT/PC					
	Subcritical, natural gas/coal (33/67)	41	121	4	125	13
	Supercritical, natural gas/coal (33/67)	49	161	3	164	36

* Gasifier + PFBC.
† Converted from HHV using a conversion factor of 1.04. Emission factor for reference coal: 25.2 gC/MJ(LHV); natural gas: 15.3 gC/MJ; and oil: 20.0 gC/MJ.

2

PROPERTIES
OF COAL

2.1
Formation of Coal

Coal is produced the process of coalification, which takes place over a period of millions of years as shown in Table 2.1. The level that a coal has reached in this coalification series is termed its rank, and coal can be classified in terms of rank, or it can be classified into different types that reflect the differences in plant material it contains, or it can be classified by grade that is associated with the range of impurity, such as ash-forming minerals, that it contains. There is considerable interest in the use of these classification schemes to predict the combustion behavior of different coals from all over the world.

Vegetable Origin of Coal

Coal can be regarded as an organic sedimentary rock. It consists of carbon, hydrogen, oxygen, and minor proportions of nitrogen and sulfur. The highly variable proportion of inorganic material always found in coal is present largely as a mixture and not in chemical combination with the organic material.

Although coal is normally opaque, it can be ground into thin, transparent slices. In this form it appears not black but predominantly yellow or reddish-orange in color in transmitted light (i.e., absorption colors). Microscopic exami-

TABLE 2.1
GEOLOGICAL AGE OF COAL AND TYPICAL VEGETATION (After Macrae, 1996)

Millions of years

Period	Total	Geological period	Era	Types of dominant vegetation	Coalfields
2	2	Pleistocene	Cenozoic		
23	25	Pliocene / Miocene (Tertiary Epoch)	Cenozoic	Vegetation of modern aspect	Most European brown coals, Japan, New Zealand, Oceania
45	70	Oligocene / Eocene (Tertiary Epoch)	Cenozoic	Vegetation of modern aspect	Japan, S. Nigeria, India (NW frontier), US (Pacific coast, Northern Great Plains, Alaska), Canada (parts of)
65	135	Cretaceous	Mesozoic	Angiosperms, plane, sycamore, oak willow, fig, palm, beech, birch	US (Northern Great Plains), Canada (Brt. Col., Sask., Alberta), Japan (part of), S. Nigeria
45	180	Jurassic	Mesozoic	Incoming of angiosperms (enclosed-seed plants)	India (NW frontier), W. China, New Zealand (south), Australia (part of), Georgia, most of Asiatic USSR
45	225	Triassic	Mesozoic	Gymnosperms (naked-seed plants)	
45	270	Permian	Palaeozoic	Conifers in upper part, marked developments of new and extinction of old forms	Kusnetz and Minusinsk (Asiatic USSR)
80	350	Carboniferous	Palaeozoic	Vascular cryptograms (spore-bearing plants). Calamites, spenophyllum, lepidodendra, sigillaria, ferns, pteridosperms, cordaites	Great Britain, Westphalia, Poland, Silesia, Saar, Donetz, Moscow Russia. Belgium, N. France, Holland, Canada (Nova Scotia), US (eastern and interior states), E. India, E. China, S. Africa (Rhodesia), Australia (part of)

nation of such thin sections reveals traces of the original vegetable structure; cell-structure can be seen and parts of plants can be identified. That is, coal is nonhomogeneous as illustrated by the photomicrograph shown in Figure 2.1. Similar structure can be seen when a polished coal surface is examined microscopically, although when an opaque polished block is studied it appears black and white. Figure 2.2 is a typical photomicrograph of a piece of coal and shows a cell structure. The types of vegetation that formed coal can often be identified by fragments recognized microscopically in the coal itself or by fossilized specimens big enough to be examined by the unaided eye. The evidence shows that the younger coal seams were formed from trees and other plants, while the Carboniferous Period coals were made from gigantic forms of more primitive plants such as tree ferns, clubmosses, and others. The types of vegetation that are known to have contributed to the coal seams of the several geological periods and the regions in which these coals are found are presented in Table 2.1.

Manner of Occurrence

Coal seams occur in two main types of deposit: the first is a sequence of coal seams (possibly as many as 20 or 30 in 500 or 1000 m thickness of strata) each of great extent and moderate thickness. The thickest individual seams in this type of deposit, or coalfield, seldom exceed 4 m. For example, in the UK the average thickness of a coal seam is about 1 m. These seams were deposited in the Devonian Period (350–400 million years ago) when the flora consisted of lycopods and small ferns submerged in shallow lagoons. In this case the thin coaly sections are separated by sequences of rock (shales, sandstones, limestones).

Fuller descriptions of coal petrology are given in numerous text books (e.g., Stach et al., 1982, van Krevelen, 1993).

In the Lower Carboniferous Period (310–280 million years ago) the appearance of conifers in thick forest swamps resulted in the deposition of thick economical seams. In North America this age is referred to as the "bituminous coal period."

Formation of Coal Seams

The main factor in the formation of coal and of coalfields has been the accumulation and partial decay of vast quantities of woody material to produce peat. Peat is the precursor to coal. It forms in swamps and marshes in areas where the climatic conditions favor rapid plant growth. The rate of subsidence in the swamp must be similar to the rate of flora growth so that accumulation is favored.

Figure 2.1   Transmission microscopic view of a coal sample showing nonuniform structure.

Figure 2.2   Cell structure shown in coal.

Two mechanisms have been suggested that favor peat formation and the accumulation of the enormous quantities of vegetable matter involved. The "in situ" mechanism postulates the growth of extensive forests in swampy ground (autochthonous swamps). The trees and other types of vegetation died and lay where they fell (in situ). In the course of geologic time the very slow sinking of the site allowed a continuous great thickness of decaying woody material to accumulate on the growing site. This, in due course, was buried under silt when the forest was submerged by abrupt subsidence and the incursion of silt-laden water. In most of the large coalfields of the Northern Hemisphere the in situ mechanism is the only sensible explanation of the occurrence of the very extensive, roughly parallel, and generally uniform coal seams of low inorganic content.

The "drift" mechanism postulates vegetation to have been carried ("drifted") by moving water (e.g., a great river system) from the growing area to a remote shallow-water site in which it accumulated and was eventually submerged and buried (allochthonous swamps). Allochthonous swamps are, in general, too rich in mineral matter, spores, and pollen to form economic coal deposits.

Coalification: Conversion of Wood to Coal

In the conversion of wood to coal, a first stage is envisaged in which vegetation decays through the agency of microorganisms (bacteria and fungi) in nearly stagnant, or slowly moving, water in great coastal swamps or in shallow, inland depressions of large area. The result is the formation of a peatlike material. It is evident that if the deposition site had been dry, instead of being waterlogged, the woody material would have decayed completely away, with the formation, principally, of CO_2 and H_2O. The presence of water retards decay by preventing the completely free access of oxygen to the decay-producing organisms.

The nature of the peatlike product of decay can be estimated from the fact that vegetable matter consists of three main types of material: cellulose, lignins, and plant proteins. Cellulose is a carbohydrate easily hydrolyzed to various extents. Plant proteins are essentially nitrogenous substances containing, frequently, sulfur and phosphorus. They are readily hydrolyzed to amino acids. The lignins are related to cellulose but are distinctive in having a benzenoid structure and are not easily hydrolyzed to simple substances. The first stages of plant decay may reasonably be regarded as occurring in a swamp environment, under conditions of reduced oxygen access, by the agency of aerobic bacteria and microfungi. Cellulose is decomposed, partly to CO_2 and H_2O and partly to a colloidal oxidation product named oxycellulose. Lignin is only slightly affected by the conditions, forming a partially hydrolyzed lignin colloidal material. The plant proteins yield amino acids. The first stages of plant decay are thus bacterially assisted partial oxidation and hydrolysis that degrades the cellulose, lignins, and proteins to colloidal products that can react to create colloidal aggregates in the swamp. In the course of this process, maceration of the woody material must have occurred, and the colloidal solutions permeated decaying woody fragments of various sizes that had reached a less advanced stage of decomposition. This phenomenon helped to preserve, in many cases, the biological structure of these fragments throughout the whole complex process of coalification.

The peatlike material may be regarded as essentially a hydrosol that in time became a hydrogel. Burial of the peat under a thick layer of silt stopped plant growth and eventually stopped bacterial action. Initially, following the subsidence and cover, the peat experienced further anaerobic bacterial attack. Eventually all bacterial action must have ceased when the silt accumulation prevented the removal of decay products toxic to bacteria through dissolution in water or by other means.

The gradually increasing weight of the accumulating inorganic layer above the buried peat caused consolidation of the peat. This pressure effect increased as successive layers of deposits were built up in the course of geologic time. The effect of pressure resulting from the overburden (overlaying layers) and also, in the course of geologic time, of other pressure effects and of temperature elevation, both resulting from movements of the earth's crust, caused further profound changes in the buried peat. By these means a whole series of types of coal was formed, each member of the series representing a different extent of metamorphosis of the peat deposit. These stages in the conversion of wood to coal are represented in Table 2.2. The most highly changed material, that is the final member of the series of coals formed from peat, is anthracite.

The different types of coal that are clearly recognizable by their different properties and appearance can be arranged in order of their increasing metamorphosis from the original peat material:

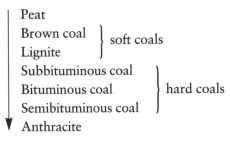

TABLE 2.2
SCHEMATIC DIAGRAM OF COAL GENESIS

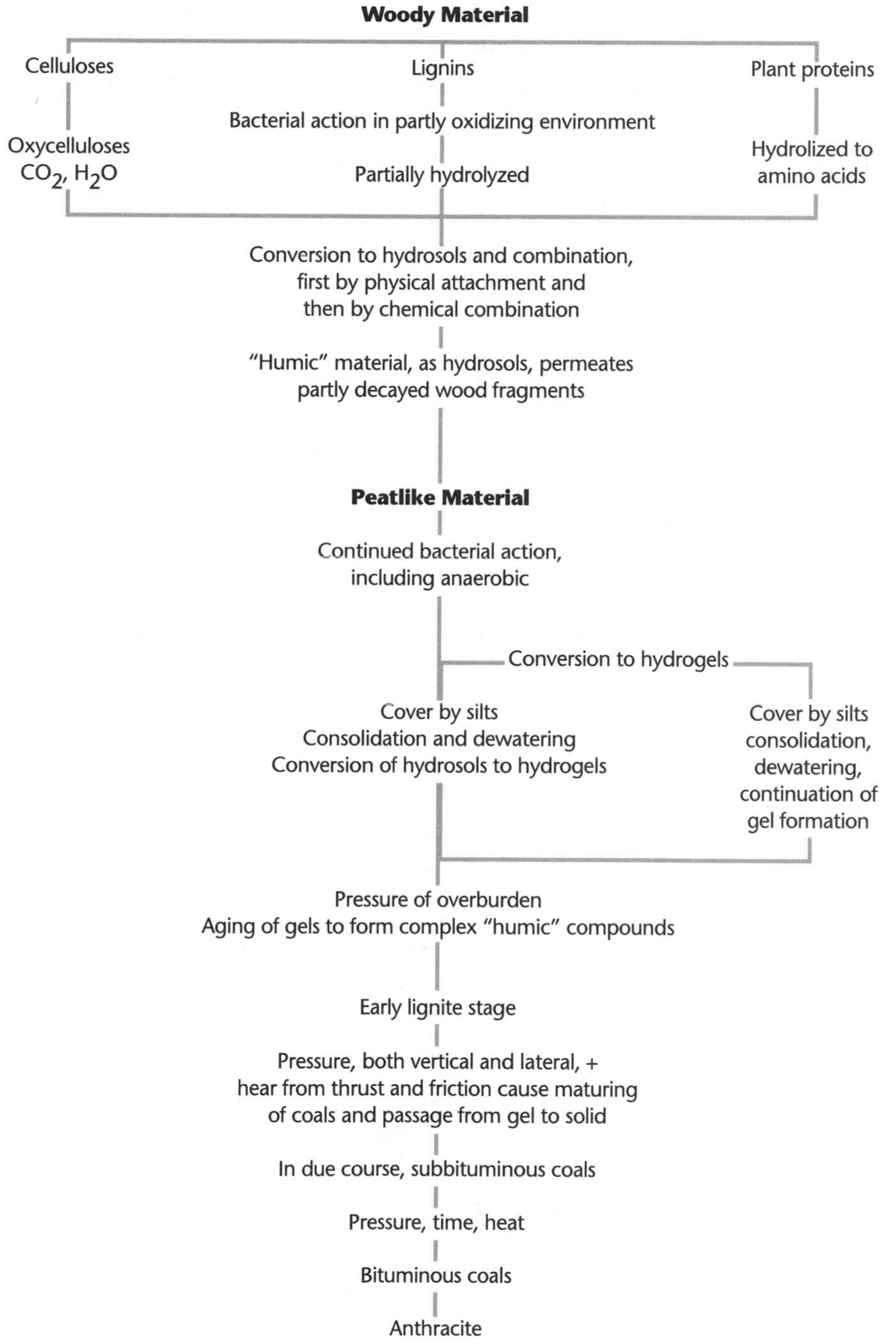

Woody Material

Celluloses	Lignins	Plant proteins
Oxycelluloses CO_2, H_2O	Bacterial action in partly oxidizing environment / Partially hydrolyzed	Hydrolized to amino acids

Conversion to hydrosols and combination,
first by physical attachment and
then by chemical combination

"Humic" material, as hydrosols, permeates
partly decayed wood fragments

Peatlike Material

Continued bacterial action,
including anaerobic

Conversion to hydrogels

Cover by silts	Cover by silts
Consolidation and dewatering	consolidation,
Conversion of hydrosols to hydrogels	dewatering,
	continuation of
	gel formation

Pressure of overburden
Aging of gels to form complex "humic" compounds

Early lignite stage

Pressure, both vertical and lateral, +
hear from thrust and friction cause maturing
of coals and passage from gel to solid

In due course, subbituminous coals

Pressure, time, heat

Bituminous coals

Anthracite

Each member of this series represents a greater degree of maturity than the preceding one, and this degree of maturity is referred to as the rank of the coal. A coal of a certain level of maturity, or degree of metamorphosis from the peat, is said to be of a certain rank. Subbituminous coal is of higher rank than brown coal, lignite is of lower rank than bituminous coal, anthracite is of the highest coal rank.

During the coalification process, the chemical nature of the organic deposit changes as summarized in Table 2.3.

TABLE 2.3
ANALYSES IN THE COALIFICATION SERIES

Material	Wood	Peat	Lignite	Subbit. coal	Bit. semibit. coal	Anthracite
Moisture (material as found)	30–60	90 +	20–40	10–20	13–1	2–3.5
Moisture (air-dried material)	10–15	20–25	15–25	10–20	13–1	2–3.5
Dry, ash-free material:						
Carbon	50	55–65	65–73	73–78	78–92	92–96
Hydrogen	6.0	5.5	4.5	6.0	5.3	2.5
Oxygen	43	32	21	16	8	4

2.2
Characterization of Coal

Coal is very heterogeneous in nature. For commercial coal applications it is essential that the chemical composition, calorific value, and in the case of some combustion gasification processes, the swelling properties be fully characterized and understood.

In general, the laboratory tests applied to characterize coal and to assess its properties are proximate analysis, ultimate analysis, and various coke swelling index experiments to clarify its properties.

These tests are described in detail in a number of books and publications (Frost et al., 1978, van Krevelen, 1993, Carpenter, 1988, Carpenter and Skorupska, 1993). Indeed, some of them are so very empirical in their character that apparently minor variations in procedure cause important variations in the quantitative results measured. It is not enough to be competent at making a test; it is necessary to be able to interpret the merits and limitations of the test results and to be able to interpret them in a wider context than that of the analytical laboratory. Detailed, authoritative instructions for making these tests are given in the ASTM, ISO, and BS* methods, and are used globally to specify coal for technical purposes and for commercial contracts. These are outlined briefly next.

Proximate Analysis

This provides data for a first general assessment of a coal's quality and type. Assessed are moisture, ash, and volatile matter.

Moisture   Moisture in coal tends to reduce its heat of combustion and hence it is of considerable commercial interest. In general, the moisture content decreases with increasing rank, reaching a minimum for the bituminous coals and increasing slightly for the anthracites, where moisture can adsorb in the microporous structure. This is indicated in Table 2.3.

Coal is a porous carbonaceous material and contains considerable amounts of water. Moreover, the amount of water held in this way in apparently dry coal varies even at constant temperature in response to variations in the humidity of the surrounding atmosphere. Because the variation can be large, materials of this kind cannot easily be weighed accurately for analysis unless they have first been allowed to achieve equilibrium with the humidity of the laboratory atmosphere. After these solid fuels have been allowed to reach this equilibrium condition preparatory to analysis, they are said to have been "air-dried." The moisture-holding capacity of coal at 30°C and 97–99% humidity measures the equilibrium moisture content (as described by the standard methods ASTM D1412, BS1016 part 21). The bed moisture can then be determined by proximate analysis (ASTM D3173) as outlined below.

Moisture in the air-dried sample is determined by heating a weighed quantity (1 or 2 g) of –72 BS analysis sample at 105°C in an inert atmosphere (oxygen-free nitrogen) to constant weight, $M''\%$. This is generally achieved in 75–90 minutes. The loss in weight is regarded as equal to the moisture content. The atmosphere in the drying oven must be quite free from oxygen; otherwise, oxygen will combine with the coal and the loss in weight on drying will not be equal to

---

*ASTM: American Society for Testing Materials; IOS: International Standards Organization; BS: British Standards, London.

the moisture content but to the moisture content less the oxygen fixed. This fixation of oxygen is especially active with the lower rank bituminous and nonbi-
tuminous coals and is seen to occur in connection with coal storage.

If the loss in weight of the bulk sample after air-drying has been measured,
the total moisture content of the coal as sampled can be calculated thus:

Weight of bulk sample as sampled a
Weight of bulk sample air-dried b
Loss on air-drying $(a-b)$
Equilibrium moisture lost, % $[(a-b)/a] \times 100 = M'\%$

Moisture determined on *air-dried sample* by proximate analysis, $M''\%$

Total moisture in *coal as sampled* $= M' + [M''(100 - M')/100] = M^T\%$

To distinguish the two moisture values, the moisture content of the bulk as sampled is referred to as "total moisture" and that of the air-dried sample is often
called "inherent moisture." It is the inherent moisture content that displays the
systematic variation with change in rank.

Ash   Ash is determined by heating a weighed quantity of sample to $825\pm25°C$
in a furnace until all the organic matter has been burned away (ASTM D3174,
BS1016 part 104). The furnace must have a free but gentle current of air
through it to ensure fully oxidizing conditions. The residue of inorganic matter
is weighed as ash.

The ash is derived from the mineral matter in the coal but did not exist in the
coal in the form in which it weighed in the experiment. The mineral matter in
coal consists of hydrated aluminum silicates, iron pyrites, calcium and magnesium carbonates, alkali chlorides, plus various other inorganic constituents in
trace amounts. When coal is burned, as in this test, the mineral matter decomposes, with the production of dehydrated oxides plus small amounts of sulfates.
The sulfates are formed by the fixation of some of the SO_3 produced in the oxidation of the coal (which itself always contains sulfur) and in the oxidation of
iron pyrites, the SO_3 being fixed by basic oxides (e.g., CaO derived from
$CaCO_3$ in the mineral matter). The weight of ash is less than that of the mineral
matter in the original coal. The proportion of mineral matter in most bituminous coals can be calculated approximately from the ash content by multiplying
by 1.15: that is, mineral matter = 1.15 × ash.

Moisture and ash (or mineral matter) are incombustible materials (with the
exception of iron pyrites) and are of no value to a consumer. They are referred

to as the inert material in the fuel. The value of a coal is basically greater the lower the proportion of inert material in it. In this connection the relationship between measured ash value and original mineral matter is important in industrial practice. For example, two coals may have the same value for total inerts as measured by moisture and ash values:

Moisture: 5%   10%

Ash: 10%   5%

Total inerts: 15%   15%

According to the analysis these coals contain the same amount of inert material (15%) and the same amount of useful combustible material (85%). This is not true because if the ash values are converted to mineral matter contents, the coals are seen to have

Moisture: 5%   10%

Mineral matter: 11.5%   5.75%

Total inerts: 16.5%   15.75%

One coal has almost 1% more inert material, i.e., 1% less combustible matter, than the other. If the coals are available at the same price per ton, one is a better value than the other in terms of combustible matter bought.

As a consequence, coal analyses (proximate and ultimate) are often reported on a dry, ash-free (daf) or a dry, mineral-matter-free (dmmf) basis:

$$\text{Wt\% C(dmmf)} = \frac{\text{measured, wt\% C}}{100 - \text{moisture, wt\%} - \text{wt\% MM}} \times 100$$

or on a dry, ash-free (daf) basis:

$$\text{Wt\% C(daf)} = \frac{\text{measured, wt\% C}}{100 - \text{moisture, wt\%} - \text{ash, wt\%}} \times 100$$

When the moisture and ash proportions have thus been disclosed, the analysis is said to have been converted to the "dry, ash-free basis." This is a very common type of calculation when the possible suitability of different coals for an industrial purpose is being considered.

Volatile matter   Volatile matter is the mixture of vapors and gases released during the pyrolysis of coal. The main constituents are carbon dioxide, carbon monoxide, water, and hydrocarbon species, including tars evolved during the decomposition of the coal macromolecular structure. Volatile matter, in general, decreases with increasing coal rank, although the amount of volatile matter released for a given coal is dependent on experimental conditions.

Volatile matter is determined by heating coal in a silica crucible for 7 min at 950°C±20°C in an electric furnace from which oxygen is excluded (ASTM D3175, BS1016 part 104). The loss in weight is equal to the volatile matter evolved as the result of the decomposition of the coal + the moisture that existed as such in the coal and was measured in the moisture test. The former is reported in the analysis as volatile matter less moisture, the measured moisture content being subtracted from the total weight observed in the volatile matter test.

Fixed carbon   The proximate analysis at this stage shows percentages by weight of moisture, volatile matter less moisture, and ash. It is completed by adding an item representing the solid carbonaceous residue from the volatile matter test, i.e., the coke residue less the ash. This is known as the "fixed carbon" (ASTM D3172). It brings the total of the proximate analysis automatically to 100.0%. Fixed carbon is the difference determined by

100 – (% moisture + % volatile matter less moisture + % ash)

The complete proximate analysis is moisture, volatile matter less moisture, fixed carbon, and ash.

The analysis has been made on air-dried coal that must be stated in reporting the analysis. From this analysis it is possible to calculate the ash, volatile matter, and fixed carbon values of the coal at any known moisture content; i.e., if the total moisture content of the original bulk sample has been determined by the proximate analysis of the air-dried sample, it can be converted by simple proportion to the proximate analysis of the coal as sampled. Thus, let M = moisture in air-dried coal, M^T = total moisture in original coal, VM = volatile matter less moisture in air-dried coal, and A = ash in air-dried coal. Then, volatile matter less moisture (in coal as sampled) is

$$\frac{VM(100 - M^T)}{100 - M}$$

Similar calculations are made with the fixed carbon and ash values.

Volatile Matter and Caking Power

The volatile matter value on the dry, ash-free basis decreases progressively with increasing rank of coal. So also do other properties. Caking power has been seen to increase to a maximum with increasing rank of coal and to decrease as the rank increases beyond a critical level. Consequently, the volatile matter value on the dry, ash-free basis gives a general indication of the caking properties of a coal.

Ultimate analysis The ultimate, or elementary, analysis is concerned with the proportions of carbon, hydrogen, nitrogen, and sulfur in coal on a wt% basis. The results are usually reported on a daf or dmmf basis.

Carbon and hydrogen Carbon and hydrogen are determined in the conventional combustion manner used with most organic compounds, but to set standards (ASTM D3178, BS1016 part 106). A small quantity (0.2 g) of the −72 BS air-dried coal is burned in a stream of oxygen in a resistance glass tube at 800°C. The tube contains various reagents. A part of the tube is packed with copper oxide and is also heated to 800°C. The gaseous combustion products pass through the copper oxide packing, which helps to complete the oxidation. Sulfur oxides and chlorine are removed by passing the fully oxidized gases through a packing of silver gauze or lead chromate and silver gauze, respectively. The gaseous combustion products that finally emerge from the combustion tube consist of H_2O (vapor) + CO_2 + excess oxygen. The water vapor, derived from the moisture in the air-dried coal and from the oxidation of the hydrogen in the coal substance, is absorbed in a weighed vessel containing a powerful, solid desiccant, such as magnesium perchlorate. The CO_2 evolved from the decomposition of carbonates and by the oxidation of the carbon in the coal substance is absorbed. This type of analysis can also be undertaken automatically by gas chromatographic based instruments.

Nitrogen The special method used for the determination of nitrogen in coal is the Kjeldahl method (ASTM D3179-89, BS1016). In this, 1 g of the air-dried sample is heated in a special, resistance glass flask with boiling, concentrated sulfuric acid until all the organic matter has been oxidized and mostly eliminated as CO_2 and H_2O. Almost all the nitrogen in the coal is converted to $(NH_4)_2SO_4$. An oxidation catalyst, commonly selenium powder, is added and the ammonia evolved is absorbed in standard acid and titrated. Coals usually contain between 1 and 2% of nitrogen.

Sulfur   Sulfur occurs in three modes of association, or forms: (1) chemically combined with the carbon of the pure coal; in this form it is known as "organic sulfur"; (2) in combination with iron as iron pyrites, FeS_2; this is the "pyritic sulfur"; and (3) as various sulfates (e.g., of Ca, Mg, Fe). Sulfur combined as sulfate (the "sulfate sulfur") is usually present in only trace quantities.

Usually only the total sulfur is determined, but methods are available for ascertaining the amount present in each of the three forms (BS1016m 106.4.2 1996, 106.5; ISO 157: 1996). A simple, standard method of determining the total sulfur present is the Eschka method. In this, the air-dried coal is mixed with the Eschka ($MgO + Na_2CO_3$) mixture, and the mixture is heated until all the coal has been oxidized. Oxides of sulfur from the combustion of the coal and of any pyrites present react with the alkaline Eschka mixture (containing $BaCl_2$) to form sulfates and sulfites. When oxidation is complete, the residue is mixed with water, and any sulfites present are oxidized to sulfates by the addition of a little bromine-water. The barium sulfate is formed and weighed. The total amount of sulfur in the coal on a wt% basis is calculated. Where the amounts of sulfur present as pyrites and as sulfates are known, the organic sulfur is found by subtracting these from the total sulfur.

Sometimes reference is made in an ultimate analysis to "combustible sulfur." This is a difference figure found by determining the sulfur fixed in the ash when the coal is burned in air and subtracting this from the total sulfur. The amount of sulfur fixed in the coal ash is related to the method used to preparing the ash sample for analysis and is often erroneously referred to as "incombustible sulfur."

Because no truly reliable method is available for its determination, the oxygen content of coal is expressed as a difference figure in the ultimate analysis and therefore includes the balance of the analytical errors. It is recorded as oxygen + errors.

Coal-swelling Tests

Certain coals (especially bituminous) swell on heating. This begins with softening at 350°C and is followed by fluidity and swelling that coincides with volatile release, and resolidification at about 500°C to form semicoke. Further heating produces coke. Because of its use in coke formation, a number of tests exist to characterize the coking properties of coal: Gray-King coke type (BS1016, 107.2 1991), the free-swelling index, and the Roga index number [(ISO 335-1994 (E)].

The Gray-King coke test involves heating finely ground coal in a horizontal, sealed, cylindrical retort for 1 hour. The appearance of the resulting coke is compared to standards labeled A to G. A to C are powdered, granular products with little cohesion; D to F are fused, but reduced in volume. Coals that have equal or greater volume upon coking are labeled G-type, with subscripts that are determined from the volume of inert electrode carbon that must be mixed with the coal in order for the product to occupy the same volume as the original sample.

Calorific Value

The calorific value is the amount of heat that can be released when coal is burned. A ground sample of coal is burned with oxygen in a bomb calorimeter under standard conditions. The gross calorific value (GCV) is calculated; this is often termed the high heating value (HHV). The net or lower value (LCV or LHV) is obtained by subtracting the latent heat of evaporation of water. Generally, calorific value increases with rank as previously described and can be estimated from the proximate analysis of the coal, e.g.,: by the Dulong formula: CV = 144.4 (%C) + 610.2 (%H) – 65.9 (%O) 0.39 (%O)$_2$, where %C, %H, %N, and %O are the respective C, H, N, and O contents of the coal on a daf basis.

Particle Sizing

Clearly, this technique must be capable of covering a number of particle-size ranges. For larger particles a sieve technique is used, and this involves the separation of a representative sample of coal into size fractions that are then weighed.

If dry sieving is used, the coal is dried and placed on the sieve with the largest aperture and the coal passes through a series of sieves by shaking. If wet slurry is used, the sample is placed on the largest aperture and the fines are washed through and collected. Particles with sizes below 63 μm and down to 2.5 μm are usually measured by means of a Coulter counter.

Thermal Gravimetry

This is now becoming a routine analysis method using a thermal analyzer. A number of measurements can be undertaken. For proximate analysis the temperature of the coal sample is slowly increased (50°/min) from 25–900°C and the mass loss measured as shown in Figure 2.3. The ash content is measured by changing the N$_2$ flow to O$_2$ at 900°C. Further details are given in Appendix 1. Details of the burning profiles, the mass loss under oxidizing conditions, can also be obtained and again details are given in Appendix 1. Examples of burn-

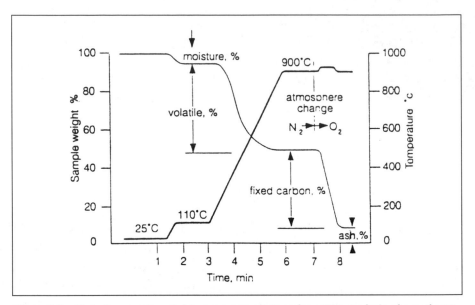

Figure 2.7 Plot of a sample weight versus temperature for a TGA analysis of a coal sample. The sample is first heated in N_2 gas to measure the moisture and volatile contents, then switched to O_2 gas to give the fixed carbon and ash.

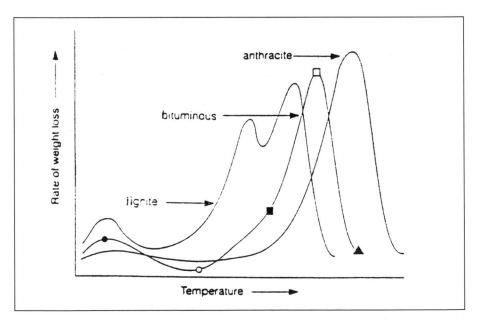

Figure 2.4 TGA diagnostic of different coals. The rate of heat loss is plotted against temperature.

ing profiles are given in Figure 2.4. Finally, details can be obtained on pyrolysis where the sample is heated under nitrogen at a constant rate (10–30°C/min) and the products examined by FTIR [see below and information obtained on pyrolysis (Figure 2.5)]. This is considered further in Chapter 4.

Reflectance

This forms part of the petrographic examination of a coal and gives not only the major maceral types but also the average vitrinite reflectance. Here a reflected-light microscope fitted with a photometer is used and the measurements made on a block of coal. Measurements are made of the microlithotype, the carbominerite, and the minerite and of the reflectance over a number of intervals.

Infrared Spectra

The development of Fourier transform infrared (FTIR) technology has transformed many aspects of coal analysis. A typical spectrum is shown in Figure 2.6, where the band assignations are shown. In principle it is possible to estimate the aromatic and aliphatic contributions, the influence and amount of mineral matter, and information on phenolic and sulfur groups (Mullins et al., 1993).

Ash Fusibility

This test gives information on the melting characteristics of the ash produced from a coal. A test piece, shaped like a cube, is heated until changes in shape occur and that temperature, the ash fusion temperature, is noted. Coals with low ash fusion temperatures (1000–1200°C) form a clinker. Generally, a minimum ash fusion temperature of 1300°C is required for the safe operation of mechanical stokers (Merrick, 1984). A number of other tests are available for ash, including complete chemical analysis.

Other Chemical Methods

These involve methods for chlorine and fluorine, sulfur by infrared spectroscopy, mercury, cadmium, arsenic, selenium, sodium and potassium.

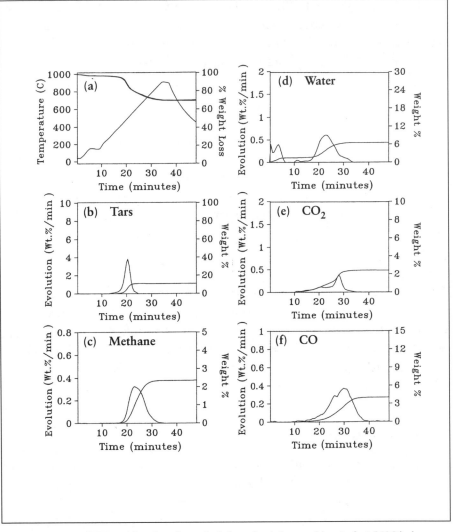

Figure 2.5   TGA/FTIR study of coal pyrolysis koonfonteine coal heated at 30°C/min, 1998 (courtesy of AFR).

Other Physical Methods

A number of other test methods are available to measure density, grindability (the Hardgrove index and abrasion index), and the surface area of coals and chars. Details of these tests are given in standard textbooks.

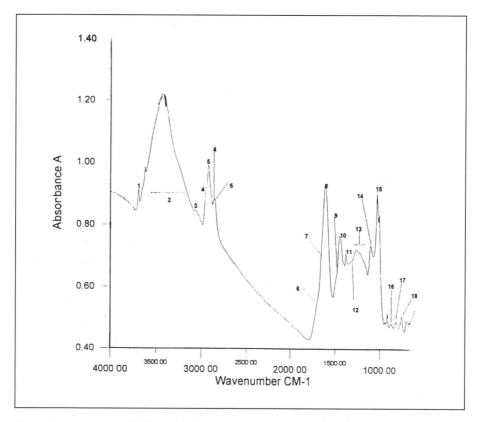

Figure 2.6 Annotated FTIR spectrum of a coal sample (Markham Main coal). 1. Crystalline bonded water; 2. hydrogen from water, alcohols, acids, and amines; 3. aromatic hydrogen; 4. CH$_3$; 5. Ch$_2$; 6. Casbonyl (C=O); 7. conjugated carbonyl (C=O); 8. amide, water, and phenols; 9. carbonates; 10. –CH3; 11. –GH3; 12. not identified peak, present only in exinite; 13. C-O-C ester and C-O phenols; 14. silicon dioxide and clay; 15. clay, ether, and alcohol; 16. aromatic hydrogen with 1 adj. H; 17. aromatic hydrogen with 3 adj. H; 18. aromatic hydrogen with 4 adj. H.

2.3
Coal Characterization and Classification

Coal Classification

The systematic variation of the properties of coal with rank, and the possibility of measuring these properties by laboratory tests so as to assess the suitability of coals for industrial purposes, has led to the creation of a number of classification systems. Each system is intended to allow many properties of coal to be deduced from a knowledge of two or three easily measured features, or to allow the possible industrial outlets for a coal to be decided from a knowledge of a limited range of properties. The main categories for classifying coal are

1. The broad classification into major classes of coal, i.e., lignite, subbituminous, etc., used in a broad descriptive way, mainly the description of coal reserves, etc. (cf. Chapter 1).
2. Classification based on coal type, i.e., the origin of the coal, coal petrology.
3. Classification based on coal rank. This is perhaps the most useful for the commercial use of coal. A number of different systems have been devised to provide a description of a coal for utilization purposes.

Classification into Main Coal Classes

Brown coals are dark brown in color and have a marked "woody" or fibrous structure. As mined, they contain 40–60% of moisture. Their carbon content (dry, ash-free basis) is low, being 60–65%. Lignites are closely related to brown coals. Indeed, the names tend to be somewhat loosely applied so that at times confusion arises. Properly, lignites are black in color. They contain 20–40% of moisture, as mined, and 65–73% of carbon (dry, ash-free basis). On the same basis their calorific value is around the low value of 25 MJ/kg (daf) and they yield 40–60% of volatile matter on carbonization.

Both brown coals and lignites are quite devoid of caking power. A major utilization difficulty with both lignites and brown coals is their high moisture content. When air-dried, they disintegrate to powder, which is convenient for pf combustion in power generation. They oxidize readily and if totally dried tend to ignite at relatively low temperatures during the drying operation. They can be briquetted after partial drying. Successful briquettes can be made either without the addition of an added binding material or with the help of additions of pitch or tar.

Subbituminous coals are intermediate in rank between mature lignites and the hard, bituminous coals. They usually have 10–20% of moisture when air-dried and, on the pure coal basis, 73% to around 78 or 79% of carbon. On carbonization they are quite noncaking but yield some 45% of volatile decomposition products. Their hydrogen content is, relatively, high and may reach 6%, whereas the organic matter of mature lignite has substantially less than 5% of hydrogen.

Bituminous coals constitute the bulk of the so-called "hard" coals, and are also the principal type of coal. The moisture content (air-dried basis) of this type of coal varies widely according to the level of rank. Commonly, the lowest rank bituminous coals contain some 13% of moisture (air-dried basis), while the highest rank varieties of this class of coal have approximately 1% of moisture. The yield of volatile matter on carbonization also is very sensitive to changes in rank. The range of carbon contents (daf) is from 78–90%, and in response to the same increase in rank the yield of volatile matter falls from 40–20% (daf). The calorific value of these coals is high and increases with increasing rank from 32–37 MJ/kg (daf). The lower rank bituminous coals (lower carbon content) are noncaking, but the phenomenon of fusibility appears for the first time among coals in this class. This rheological behavior on carbonization is first seen in, for example, British coal seams by the bright coals of 83% carbon content. Caking power in this class of coals increases with increasing rank, and the bituminous class includes the strongest coking coals known.

Semibituminous coals can be regarded as a transition class between the bituminous and anthracite coals. They are all of low inherent moisture content (0.7–1.5%), high in carbon (90–93%, daf) and low in volatile matter yield (20–10%). The term "inherent moisture" is a synonym for "moisture on the air-dried basis." This class also contains coals covering a wide range of caking power from strongly caking to noncaking varieties, but in the semibituminous class caking power decreases with increasing rank.

Anthracites are now very much less widely distributed than are the bituminous coals; many coalfields based on anthracite are now exhausted.

Classification Based on Coal Petrology

Because of the way that coal is synthesised from plant fragments and is associated with clays, rocks, and other mineral impurities, it is a very heterogeneous material. Inspection of any lump of coal shows that it consists of a large number of bands, the lithotypes, of varying thickness and appearance. This is most apparent in coals of bituminous rank. When a coal seam is examined in situ in a coal mine, the bands are seen to be horizontal. They represent successive layers, or depositions, of vegetable matter in the original "peat bog." The branch of science concerned with this visible structure of coal is coal petrology. The structure may be examined visually by the unaided eye or by the optical microscope.

Two main types of banded constituent can be distinguished by the unaided eye. One is bright in appearance and the other is dull. Bright coal constitutes the main part of most seams of hard coal. Figure 2.7 is a photograph of the lithotypes of a clearly banded bituminous coal. It shows the two types of bright coal

Figure 2.7   Photograph of a lump of coal showing banded structure.

and bands of dull coal. There are, however, four lithotypes: two are bright in appearance, and two are dull. They are

Vitrain. Brilliantly glossy and vitreous in appearance. It occurs as thin bands generally 2–10 mm thick. Vitrain is brittle so that coals very rich in this component tend to disintegrate on industrial-scale handling and influence their size distribution during milling.

Clarain. Bright coal less brilliantly glossy than vitrain but with a satin luster and a clearly striated (i.e., laminated) structure. Clarain is mechanically stronger than vitrain. When carbonized, it normally yields more volatile matter than vitrain and yields a more compact coke.

Durain. Banded dull coal. It is hard and has a granular, matte surface. (Normally, durain is gray in color; exceptionally it may appear a sooty black that becomes brown when scraped.) Durain is much harder than either vitrain or clarain. Gray durains do not yield fused cokes whatever the rank of the coal seam may be. Gray durain contains more carbon and less hydrogen than the bright constituents. It also yields less volatile matter than these but has a higher calorific value.

Fusain. Powdery, dull, black. It occurs as thin patches or wedges in the bedding plane. Fusain is always entirely noncaking and tends to reduce the caking power of any bright coal in which it occurs. Compared with the other banded constituents, fusain is always high in carbon content and low in hydrogen content (both on the dry, ash-free basis).

Vitrain and clarain together make up the bulk of a normal coal seam. Durain and fusain are anomalous constituents. Note especially that the systematic variation of properties with rank is shown by the bright constituents and not by the dull ones. Therefore, in using some feature of the analysis of a coal to predict an approximation to another property, only a bright coal analysis may be used.

Gray durain and fusain tend to contain more inorganic (ash-forming) material than either vitrain or clarain.

The lithotypes contain within them microcomponents, the microlithotypes, association of individual macerals: vitrain has the main microlithotype called vitrite, durain has the main microlithotype called durite, clarain has the main microlithotype called clarite, and fusain has the main microlithotype called fusite.

The composition by the smallest components, the macerals, of the main microlithotypes are listed below:

Microlithotype group	Composition by maceral groups
Vitrite	> 95% vitrinite, < 5% exinite + inertinite
Durite	> 95% inertinite + exinite, < 5% vitrinite
Clarite	> 95% vitrinite + exinite, < 5% inertinite
Fusite	> 95% inertinite

The maceral groups, which can be measured by the reflectance under a special reflecting microscope, are related in turn to the macerals and the individual plant debris from which they were derived:

Maceral group	Maceral	Origin
Vitrinite	Colinite	Humic gels (cell cavities)
medium	Telinite	Wood, bark, stems, leaves
reflectance		(cell walls)
	Vitrodetrinite	Plant debris
Exinite (liptinite)	Sporinite	Skins of spores and pollens
(High H-content)	Cutinite	Cutine (cuticles on leaves, needles,
low reflectance		shoots, stalks, etc.)
	Resinite	Resin
	Alginite	Algae (seen in boghead coals)
	Liptodetrinite	Fragments of spores, cuticles,
		resins, algae
Inertinite	Fusinite	Carbonized wood (charcoal)
(Inert during coking, i.e.,	Semifusinite	Less well-carbonized wood
do not change upon heating)	Macrinite and micrinite	Unspecified debris
High reflectance	Sclerotinite	Fungal remains
	Inertodetrinite	Fragments of inertinite macerals

Thus, coal must be regarded as a heterogeneous material. The properties of a coal seam depend upon three principle features, namely:

1. The rank of the coal. The carbon content of the pure (i.e., dry, ash-free) bright coal is commonly taken as an index of the rank of coal, and this is further discussed later.
2. The petrographic composition.
3. The mineral matter in the coal, which is discussed next.

Mineral matter in coal   After combustion, coal ash usually contains 95% of alumina, silica, iron oxide, calcium oxide, and magnesium oxide and 5% of sodium oxide, phosphate, germanium oxide, titania, and gallium oxide. The mineral matter is distinct from ash, but mineral matter can be related to ash empirically. There are two sources, namely: (1) Inherent mineral matter arising from the inorganic constituents of the plant materials from which the coal is formed. (2) Adventitious mineral matter transported to the coal bed from a different location. This can be one of two types: syngenetic mineral matter that was laid down alongside the coal or transported into the beds before the coal was deposited, and epigenetic mineral matter that was deposited into the peat bog by descending (or ascending) in cracks and fissures.

Mineral matter and lithotypes   Mineral matter deposited by syngenetic means depends on the diffusion of mineral solutions into the peat materials, and the fixation of minerals by adsorption.

1. *Woody tissue* (origin of vitrain). Retains its texture on decomposition—little diffusion into its structure → relatively little ash.
2. *Highly decayed "mud"* (origin of durain). Permeable → high ash content.
3. *Clarain.* Intermediate between these two.
4. *Carbonized wood* (origin of fusain). Spongy texture; adsorbs solutions of mineral into its pores → highest ash content of the lithotypes.

Mineral matter deposited by epigenetic means depends on the permeability of the lithotypes: fusain > durain > clarain > vitrain. Therefore, the overall ash contents of the macerals follows the order: fusain > durain > clarain > vitrain.

Classification by rank   This method of classification is the most widely used for the marketing and export of coal. It is based on the fact that there is progressive change in composition through the coal rank series, and some typical analyses of some of the broad classes are in Table 2.3.

From the data in Table 2.3, it should be noted that the members of the peat-to-anthracite series display a steady decrease in the water content of the material with increasing rank and hence a steady increase in the content of combustible matter (dmmf) to a situation in the semibituminous coals at which the moisture content (air-dried basis) is only 1%. Further increase in rank toward the anthracites is accompanied by a reversal of this trend.

The pure coal free from water and inorganic mineral matter (dry, mineral-matter-free) shows a steady increase in its proportion of carbon and a steady decrease in its oxygen content through the whole range of increasing rank. The proportion of hydrogen decreases as the rank increases to a low value in the more mature lignites. In the interval of rank from the lignites to the subbituminous coals, the progressive change in hydrogen content is interrupted. Instead of declining, the hydrogen content of coals abruptly increases in this interval and then resumes its steady reduction to a minimum value in the anthracites.

Other properties of coals change systematically with the rank of the materials. The calorific value (dry, ash-free basis) increases steadily with an increase in rank to a maximum in the semibituminous range and then tends to decline slightly in the anthracites.

The amount of tar and gas that coals evolve when they are heated decreases as the rank increases. Thus, anthracites (highest rank coals) evolve less than 10% of their weight as volatile matter when heated and almost all is in the form of gas.

These progressive changes in composition with increasing degree of metamorphism (i.e., with increasing rank) form the basis of all classification systems based on rank and can be shown graphically by plotting the analyses of a large number of coals as a graph of one feature vs. another feature that is taken to be an index of rank. Commonly, the carbon content (daf) is used as the index of rank because of the progressive nature and wide range of its change. Equally, however, plots are made using the oxygen content (daf) or the volatile matter evolution (daf) as indexes. In this plot all the points form a narrow band. Sections of the band are occupied by points relating to the different classes of coal, i.e., lignites within one section, subbituminous coal in the next section, and so on, although there is some overlapping of the sections.

A plot of hydrogen vs. carbon is known as the coal band of a particular coalfield, a technique developed by Seyler using UK coals and often called the Seyler diagram. Figure 2.8 is such a plot of many analyses of US coals, the North

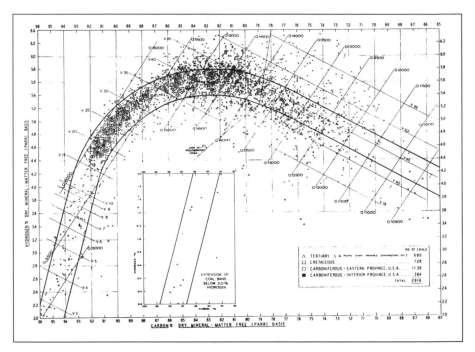

Figure 2.8   The American Coal Band: plot of hydrogen versus carbon (after Macrae,1966).

American Coal Field by Mott (Macrae, 1966). It is particularly important to appreciate that each value of the index of rank produces a substantial range of values of the hydrogen value—there is a wide spread of values. Nevertheless, the generally progressive change of hydrogen content with carbon content is clearly evident. The generality and the spread within and even around a band are found with any plot of two properties of a range of coals all taken from the same coalfield. The plot of %H (dmmf) vs. %C (dmmf), Seyler's coal chart, and a superimposed plot of calorific value and volatile matter enabled the prediction of certain coal properties from the measurement of another.

UK National Coal Board (NCB) Classification

In addition to the physical and chemical properties that have already been considered, another property is the ability to form a fused coke by the process of heating coals in the absence of air. Some coals when carbonized fuse and form a coke, other classes of coal do not form a fused coke. Instead, they leave a solid residue on carbonization of charred material that is of the same shape as the original coal (it has not fused) and is soft and easily crushed.

Because of the then large coking industry in the UK, the NCB classification was developed in the 1920s from the need to predict the caking properties of a new coal. (It was later called the British Coal Classification.) Thus, this system uses dry, ash-free volatile matter yield as a rank parameter and has a subclassification according to an empirical laboratory means of assessing caking power (the Gray-King coke test), as summarized in Table 2.4. While this classification method was based on the needs of the coal coking industry, the coal combustion industry, including the present-day pulverized coal industry, was able to use such a ranking system together with a few other tests to predict coal combustion behavior. However, in the last few years there has been a constant demand for a better classification system and better or additional coal characterization tests. The British Coal (NCB) Ranking method was devised for British-like coals only, and a number of more universal methods were devised in the US, Europe, and Australia.

US (ASTM) Coal Classification System

This system relates coal rank to its gross calorific value. It uses simple tests, proximate analysis alone for higher rank coals [> 69% C (dmmf)], and calorific value and agglomerating properties for lower rank coals [< 69% C (dmmf)]. This is summarized in Table 2.5. The fixed carbon is measured by proximate analysis and converted to a dmmf basis. The mineral matter content is calculated using the Parr formula:

$$MM = 1.08A + 0.55S$$

where MM = wt% mineral matter, A = wt% ash, and S = wt % total sulfur.

The International Coal Classification Scheme (ISO)

The ISO scheme is used in both coke production and power generation industries. For hard coals the system is similar to that of the NCB scheme. A three-digit code classifies the coal (Table 2.6). The first digit represents volatile matter content (daf) of the coal, the second represents the free-swelling index or Roga index, and the third represents the Gray-King coke type. A four-digit code is used for soft coal, and the second two digits represent the tar yield during low-

TABLE 2.4
UK NATIONAL COAL BOARD (NCB) CLASSIFICATION

* Based on the proximate analysis (volatile matter) and the Gray-King test
* Based on the proximate analysis (volatile matter) and the Gray-King test

Coal rank code		Volatile matter (% dmmf)	Gray-King coal type[†]	General description
Main class	Class			
100*		<9.1	A (non)	**Anthracites**
	101	<6.1		
	102	6.1–9.1		
200*		9.1–19.5	A–G8	**Low volatile coals**
	201a, 201b	9.1–13.5	A–C	(noncoking)
	202–204	13.6–19.5	C–G	(coking)
300*		19.6–32.0	A–G	**Medium volatile coals**
	301	19.6–32.0	>G4	(prime coking coals)
400–900		>32.0	A–G9	**High volatile coals**
400	401–402	>32.0	G9	(very strongly caking)
500	501–502	>32.0	G5–G8	(strongly caking)
600	601–602	>32.0	G1–G4	(medium caking)
700	701–702	>32.0	E–G	(weakly caking)
800	801–802	>32.0	C–D	(very weakly caking)
900	901–902	>32.0	A–B	(noncaking)

*Volatile content <19.5%—these coals are classified based on their volatile contents alone.
[†] Description of solid residue from the Gray-King test:

A	pulverent
B	breaks into powder on handling
C	coherent, but crumbles on handling
D	smaller than original coal sample, moderately hard
E	smaller than original coal sample, hard
F	slightly smaller than original coal sample, hard
G	same size as original, hard
G1	slightly increased in size, hard
G2	moderately increased in size, hard
G3	increased in size compared to original coal sample, hard
G4–G9	increased in size compared to original coal sample, hard

TABLE 2.5
US (ASTM) CLASSIFICATION

Class	Group	Fixed carbon (% dmmf)	Calorific value (Btu/lb mmmf)	Agglomerating character
I. Anthracitic	1. Meta-anthracite	≥ 98		Nonagglomerating
	2. Anthracite	92–98		
	3. Semianthracite	86–92		
II. Bituminous	1. Low volatile bituminous coal	78–86		
	2. Medium volatile bituminous coal	69–78		Commonly agglomerating
	3. High volatile A bituminous coal	< 69	≥ 14,000	
	4. High volatile B bituminous coal		13,000–14,000	
	5. High volatile C bituminous coal		11,500–13,000	Agglomerating
III. Sub-bituminous	1. Subbituminous A coal		10,500–11,500	
	2. Subbituminous B coal		9,500–10,500	Nonagglomerating
	3. Subbituminous C coal		8,300–9,500	
IV. Lignitic	1. Lignite A		6,300–8,300	
	2. Lignite B		< 6,300	

temperature carbonization. Hard coals are defined as having a gross calorific value of above 23.86 MJ/kg, and soft coals below this value.

Australian classification The Australian scheme for hard coals is based on the volatile matter, swelling index, Gray-King coke type, and ash content. Ash is an important parameter for Australian coals, which generally have high ash contents.

Codification The ECE coal committee has developed a coal codification that attempts to incorporate differences in coal type and grade as well as coal rank. It draws on information of vitrinite reflectance and reflectogram, maceral composition, swelling number, volatile matter, ash, sulfur, and gross calorific value.

TABLE 2.6
INTERNATIONAL CLASSIFICATION OF HARD COALS BY TYPE

CCDE numbers — The first figure of the code number indicates the class of the coal, determined by volatile matter content up to 33% VM and by calorific parameter above 33% VM
The second figure indicates the group of coal, determined by caking properties
The third figure indicates the sub-group, determined by coking properties

Groups (determined by caking properties)			CCDE numbers						Subgroups determined by coking properties		
	Alternative group parameters									Alternative subgroup parameters	
Group number	Free-swelling index	Roga index							Sub-group number	Dilatometer test (% dilat.)	Gray-King assay (coke type)
3	>4	>45		435	535	635			5	>140	>G8
			334	434	534	634			4	50–140	G5–G8
			333	433	533	633	733		3	0–50	G1–G4
			332 a / 332 b	432	532	632	732	832	2	<0	E–G
2	2 1/2 –4	20–45	323	423	523	623	723	823	3	0–50	G1–G4
			322	422	522	622	722	822	2	<0	E–G
			321	421	521	621	721	821	1	Contraction only	B–D

Classes (determined by volatile matter up to 33% VM and by calorific parameter above 33% VM)

Group number	1	0		Class 1	Class 2	Class 3	Class 4	Class 5	Class 6	Class 7	Class 8	Class 9	Subgroup		
	1–2	5–20			212	312	412	512	612	712	812		2	<0	E–G
					211	311	411	511	611	711	811		1	Contraction only	B–D
	0–½	0–5	A	100 (B)	200	300	400	500	600	700	800	900	0	Non-softening	A

Class number

Class number	0	1	2	3	4	5	6	7	8	9
Class parameters — Volatile matter (daf)	0–3	>3–10 (>3–>6.5 / 6.5 / 10)	≥10–14 (10)	>14–20 (>14–>16 / 16 / 20)	>20–28 (>16)	>28–33	≥33	>33	>33	>33
Gross calorific value (maf) kcal/kg (30°C, 96% humidity)	—	—	—	—	—	—	>7750	>7200–>7750 / 7750	>7200 and / 7200	>7200–>6100– / 6100 and less

As an indication, the following classes have an approximate volatile matter content of:

6: 33–41%
7: 33–44%
8: 35–50%
9: 42–50%

2.4
Coal Chemical Structure

The molecular structure of coal is a function of the chemical composition, the functional groups present, and their arrangement in space. Initially the construction of coal structure models was an interesting background aspect of coal science, but increasingly over the last few years they are starting to play an important role. The three-dimensional molecular structure of coal has been deduced from combined research into coal genesis, x-ray diffraction, and optical and other physical characterization methods, together with chemical studies such as solvolysis, hydrogenolysis, pyrolysis, and oxidation. More recently, techniques such as FTIR spectroscopy, nmr and mass spectrometries, and x-ray methods have been applied also (Given, 1960, Geerstein et al., 1982, Attar and Hendrickson, 1992).

As a result of these studies it is generally accepted that coal is composed mainly of aromatic units arranged into a three-dimensional structure by covalent cross-links and hydrogen bonding. Heteroatoms are incorporated into the aromatic units and functional groups. Coal is thought to contain smaller molecules dispersed within the macromolecular network. Thus, any model coal chemical structure represents a statistically averaged picture. The information required to construct such a model apart from the elemental composition are therefore:

1. Aromaticity and the average number of aromatic rings per "structural unit"
2. Cross-link density; their nature
3. Functional groups; their number and nature

The aromaticity can be deduced to some extent from x-ray diffraction studies. This can give information on the size of the aromatic units and their degree of alignment and stacking. Other approaches for investigating the aromatic units and cross-links include FTIR spectroscopy and ^{13}C and ^{1}H nmr studies (Fletcher et al., 1990, Smith et al., 1994). These techniques also provide information on the aliphatic cross-links.

The functional groups and heteroatoms in coal contain mainly oxygen, sulfur, and nitrogen heteroatoms as detected by many techniques, but most commonly IR spectroscopy, XPS, and XANES. Oxygen, between 1–20 wt% depending on rank, is present in the organic matrix as carboxylic acid, carbonyl and hydroxyl functional groups, and either cross-linkages, as well as heterocyclic oxygen. The organic sulfur is present mostly as alkyl sulfides, thiophenes,

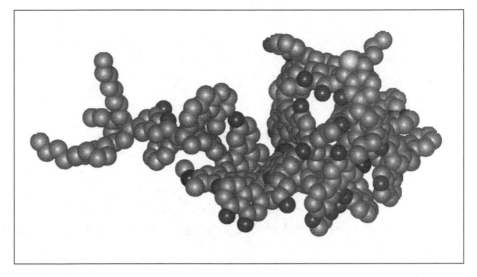

Figure 2.9 Geometrically mimimized Shinn coal model.

disulfides, and arylsulfides as determined by XPS and XANES. The nitrogen in coal is virtually exclusively organic, and typical nitrogen contents are in the range 0.7–2.1 wt%. Most of this nitrogen is pyrolic and pyridinic, the proportion of pyridinic functionality increasing with coal rank. Amines may be important in low-rank coals together with pyridone-nitrogen.

Using this kind of information, model coal structures have been devised by various authors. The earlier models were developed mainly from x-ray diffraction data and from IR and nmr spectroscopies. These models incorporated non-planar, twisted molecules that would explain the limited stacking detected by XRD and the porous nature of coal. Along with the main structural features of small aromatic units (1–2 rings) with methylene bridge linkages, functional groups such as pyrundines, quinones, carbonyl, and hydroxyl groups were proposed. Pitt's models for 80% C and 90% C vitrain were similar, but with the aid of more detailed nmr data, it was proposed that the links between aromatic units contained 9,10 dihydrophenanthrene structures. These are detailed in Van Krevelin's book (1993).

Subsequent models were influenced by the structural concepts of polymer science and contained cross-linked networks, for example, the model proposed by Wiser (1975). In the 1980s, two models were described by Spiro and Kosky (1982) and Shinn (1984). They agree with experimentally determined parameters—ultimate analysis, aromaticity, size of aromatic ring clusters (structural units), and functional groups—and can be described as "reactive models" in that they allow the interpretation of a reaction process. In Spiro's model, the process of coalification from 75 to 96% C is represented by three "space-filling" models. This was the first demonstration of the importance of steric considerations in the postulation of coal molecules. Molecular modeling techniques now make the postulation of energetically accessible structures facile. Geometry optimization and conformational analysis can eliminate energetically unfavorable structures and conformation of the various models using this method. Shinn's model was constructed from the retrosynthesis of products from liquefaction procedures. It incorporates smaller molecules dispersed within a larger polymeric structure, as shown in Figure 2.9; a geometry optimized conformation is also shown.

3

POLLUTANT FORMATION AND METHODS OF CONTROL

The combustion of coal can be undertaken by using a fixed bed of large particles on a grate, a fluidized bed of smaller particles, or the high-intensity combustion of a cloud of small pulverized coal particles (called pf or psf). Combustion can be undertaken using air, oxygen-enriched air, or pure oxygen. In general, the nature of the pollutants formed is the same, although the extent of each of the pollutants may differ depending on the combustion temperature, the combustor conditions, and the time-history profiles of the reactants and products. The process of gasification is different in that it takes place under reducing conditions, the products are different, and consequently, the nature of the environmentally unfriendly species is also different—and is treated differently. Gasification is discussed in Chapter 8.

The object of this chapter is to discuss the general way in which pollutants are produced in a flame and to outline methods for their control in situ and in the postflame gases. Details of the pollutant formation routes are outlined in a later chapter. Control methods for removing pollutants from the flue gases are considered only in outline here, but the detailed process engineering methods involved are not considered because a number of other specialized books deal with that topic (e.g., Soud and Takeshita, 1994, Speight, 1993).

The pollutants that are formed during the combustion of coal are particulate materials consisting of ash, unburned carbon and smoke, aerosols, and the gaseous pollutants carbon monoxide (CO), nitrogen oxides (NO_x), sulfur oxides (SO_x), and unburned hydrocarbons (VOC) and dioxins. The first stage of coal combustion leads to the formation of volatiles and chars. The volatiles consist of light gases, including SO_2, aromatics, and nitrogenous tars that themselves can form some of the pollutants. In addition to carbon, the chars contain ash, nitrogenous compounds, and traces of toxic metals, and as the chars burn away NO_x is released. The residual ash may contain unburned carbon that may influence the way in which the ash is utilized. The way in which these pollutants are formed is indicated in Figure 3.1, and it is clear that the amounts of pollutants formed are highly dependent upon the type of application. However, here the mechanism of pollutant formation and the principles of its control will be considered in a general sense. During the life cycle of coal utilization two other forms of environmental pollution should be noted, although they are not dealt with here. The first is that the production and supply of coal may involve

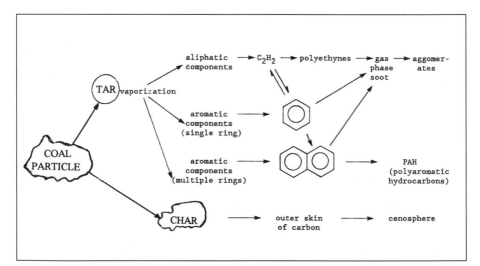

Figure 3.1   Pathways to soot, PAH, and unburned carbon. The cenosphere contains mainly solidified ash and some unburned carbon.

methane production either from mining or combustion. Second, coal combustion, in common with other fossil fuels, produces carbon dioxide that plays a significant role in the greenhouse effect, which has been discussed already and which forms an indicator of the efficiency of fuel utilization.

While it is desirable that the production of combustion-generated pollutants be minimized, it is also essential that whatever method of control is employed, the full energy potential of the fuel be realized. That is, there be, as far as possible, no reduction in the overall efficiency of the combustion process or in the utilization of the energy produced. This aspect has been particularly emphasized recently because of increased fuel costs and because of the need for conservation of fuel resources. Improvement of the overall efficiency, either directly or indirectly, automatically decreases the yields of pollutants since the total volume of combustion products is reduced.

Emissions of pollutants from coal-fired combustion equipment are subject to legislation in most countries. In the case of stationary equipment in most developed countries, this legislation was originally concerned only with the level of the emission of particulates, but now it includes emission of the acidic gases SO_x and NO_x in the case of larger plants and in some countries even relatively small plants. Some common factors are given in Appendix 1.

3.1
Formation and Control of Particulate Material

The historical perspective of coal combustion from open grates and uncontrolled furnaces was of black smoke due to the combustion of the devolatilized aromatic-containing products. In controlled combustion, the emission of smoke is very small and attention was then directed to the emission of the acid gases, SO_x and NO_x, to the emission of very small particles of coal ash, and especially to the particles of unburned carbon in the coal ash.

General Features of Smoke Formation

The particulate material found in the combustion products from coal-burning plants may be termed carbon, smoke, soot, or stack solids, depending on their size and on the particular application. In general, it consists of three groups of products as outlined in Figure. 3.1. First, it may contain smoke that is formed via a gas-phase combustion/pyrolysis process. Second, it may contain ash particles or cenospheres that are largely produced from the mineral matter present in the fuel and that may also contain small amounts of unburned carbon; this is called carbon-in-ash (or loss on ignition, LOI, although the two definitions are not the same numerically).

During the devolatilization step, the pollutants formed include gaseous hydrocarbons, hydrogen, tars, and oxygenated species such as CO. The detailed mechanism of soot formation by means of the gas-phase route is complicated, but the general features are understood. During the combustion of hydrocarbons it is generally accepted that the paraffinic hydrocarbon radicals that play an important role during the combustion process are decomposed (pyrolized) forming acetylene (ethyne), while the aromatic components form multiring structures (polyaromatic hydrocarbons). The acetylenes (ethynes) thus produced then polymerize to form polyethynes; these, together with polyaromatic rings, then form soot particles that coagulate to give the final products. The nature of the final product is dependent upon the composition of the volatiles, particularly with respect to the concentrations of aromatics, which in turn depends to some extent on the rank of the coal. The properties and the amount of the soot produced are somewhat dependent upon the residence time and the temperature, although the soots formed by all flames are remarkably similar both chemically and physically. Typically the soot product contains 90–98% (wt) carbon and only small amounts of oxygen. The particle sizes can vary widely but are usually in the range of 10–1000 nm, depending on the extent of their compilation. If aromatics are present the yield is enhanced.

The initial conversion into the particulate solid phase is known as nucleation or soot inception, and these nuclei grow by a heterogeneous process in which the soot precursors deposit on the nuclei. In the case of paraffinic volatile components, these precursors are acetylene (ethyne), and in the case of aromatic compounds and tars they may involve aromatic groups; volatiles involve both components and a general route is illustrated in Figure 3.1. Flame ions also play a part, but it is currently believed that the ionic contribution to soot formation is probably less than 10%.

Soot that is first formed in a flame reaction zone ("young" soot) may contain up to 8 wt% of hydrogen so that its composition is given approximately by the empirical formula CH (both benzene and ethyne have the same empirical formula). Longer exposure to high temperature causes the loss of most of this hydrogen so that the composition eventually shifts to C_8H, and well-established soot particles thus contain only about 1 wt% of hydrogen. The particles agglomerate together to form chains, which chains may be long or short (only a few particles agglomerate depending on the reaction conditions and particularly the reaction time). The size of soot particles emitted from coal flames is in the range 0.01–1 µm.

The individual particles that make up the chain have a diameter of about 50 nm, but they can be higher in very sooty flames. Each particle consists of a large number (10^4) of crystallites of graphitic material. These consist of 5–10 hexagonal sheets of about 100 carbon atoms lying on top of one another. The sheets are parallel to one another, but unlike graphite, they are stacked in a random way, termed a turbostratic structure, and with an intersheet spacing much greater than for graphite.

Some hydrocarbons have a greater tendency than others to form soot, and the molecular structure of the hydrocarbon is one of the major parameters determining the amount and rate of soot formation, the other parameters being the fuel/oxidant ratio, the gas temperature, and the pressure. In the soot-forming zone the reactions lead to formation of polyaromatic compounds with increasing size, and the coagulation of the heaviest of these (with molecular mass of about 1000) causes soot nucleation. The smaller species, ethyne and PAH species, are involved in surface growth of the soot particles once they are formed.

It should be noted that polyaromatic hydrocarbons (PAH) can be formed as shown in Figure 3.1. The amount and distribution of PAH isomers favors those that are thermodynamically most stable. Typical PAH compounds formed during combustion are given in Figure 3.2. Some of these are highly carcinogenic, e.g., benzo[a]pyrene (which has the acronym BaP), while others, e.g., anthracene, chrysene, are not. These, together with lighter hydrocarbons (the volatile organic compounds VOCs), may be emitted from coal flames as a group called toxic organic compounds (TOCs).

In coal-particle combustion, soot-forming conditions are always achieved (at least in part) since the region between the particle surface and the surrounding flames is always fuel-rich. Inevitably soot is always produced in this region during the combustion of the volatiles emitted, and the flame zone always exhibits a yellow luminosity. All coal flames have a yellow soot emission associated with them. The amount of soot produced in this way may be greatly reduced if the particle is burning under wake flame conditions, which approximate to premixed flames, and soot yields may be correlated with the velocity and the particle size. However, if the flames are lean enough, and well mixed enough, all the soot is burned out before the end of the combustion chamber and the flue gases contain very low levels of actual soot. Burn-out of soot tends not to occur in "open" coal fires, and this is the major source of pollutants from them.

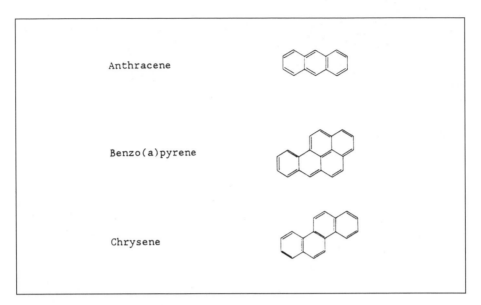

Anthracene

Benzo(a)pyrene

Chrysene

Figure 3.2   Some PAH compounds formed by the combustion of coal.

Formation of Unburned Carbon

The char formed by the devolatilization process burns away in the subsequent oxidation zone, but this may not be complete by the time it reaches cooler parts of the furnace. Cenosphere formation can occur during pf combustion. Essentially in their form the cenospheres consist of hollow inorganic spheres, but they may contain a small amount of unburned carbon. However, cenosphere-like products are formed as intermediate during the combustion process that contain substantial amounts of unburned carbon and some volatile matter; this aspect of combustion is described more fully in the next chapter.

Methods of Reducing Emissions of Particulate Materials by Controlling the Combustion Process

Generally, the formation of soot or unburned carbon can be minimized by means of the provision of good mixing resulting from turbulence, the maintenance of a high temperature, and a sufficient residence time, depending on the application. However, the scale of turbulence is also important since smoke and unburned hydrocarbons can originate from turbulent eddies that are fuel-rich.

In the case of fixed-grate combustion, this may be difficult and burn-out of the soot occurs above the bed. In the case of stationary combustion plant, the levels of excess air are kept very low but good mixing results in high flame temperatures and hence greater combustion intensity and efficiency. Because of the low excess air the yields of SO_3 are also minimized. However, as the excess air is reduced, soot formation results and this limits the reduction possible, although low levels of excess air (e.g., 0.3% O_2) may be achieved in well-mixed and well-controlled systems such as power station boilers. The operating regime in which low smoke and low SO_3 concentrations are produced is limited as illustrated in Figure 3.3. Good control is essential in such circumstances because any small reduction in the air supply would result in soot formation.

The emission of combustible particulate material can be controlled especially in the more sophisticated plants, such as power stations. The combustion conditions that can be varied are:

1. The degree of swirl or other means of mixing with air
2. The extent of recirculation
3. The combustion intensity

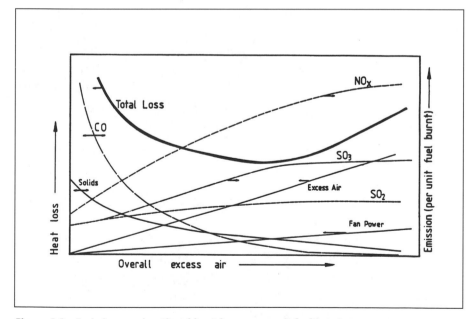

Figure 3.3   Emissions and notional heat losses per unit fuel input.

4. The quality of the ground coal, which is dependent on the method of
   milling used and the use of classifiers to remove the larger particles
5. The spatial distribution of the pf in the burner insomuch as it determines
   the local fuel/air ratio

The adjustment of any of these parameters that will make the flame tend to be-
have as a premixed gaseous flame will result in the reduction of smoke. In this
respect the key factors are the particle size, the degree of mixing, the relative
particle to gas velocities, and the gas composition.

The degree of swirl imparted to the combustion air has a considerable influence. Low levels of swirl produce considerable amounts of smoke because of poor
mixing, and increasing swirl decreases soot formation. Recirculation of combustion gases will, to a certain degree, reduce soot and unburned carbon. This can be
achieved either by swirl (internal recirculation) or by actually directing the gases
from the flue to the air intake of the burner (external recirculation). In both cases
the net result is that hot vitiated gases are mixed with the incoming air, increasing
its temperature but slightly reducing the oxygen content.

Formation of Coal Ash

Ash is the inorganic residue left after the combustion of coal, and it originates
from the inorganic materials, the mineral matter, originally present in the coal.
The most common constituents are sodium, silica, aluminum, calcium, magnesium, and iron, the first two being the most important. Typical principal mineral
matter is indicated in Table 3.1 and the compositions of typical ashes are given
in Table 3.2. Ash is left on the bed in fixed- or traveling-grate systems, and in pf
systems it is discharged from the bottom of the boiler as a powder, but in slagging (high-temperature) furnaces it is discharged as a glassy slag. The fine particles that are emitted with the flue gases are known as fly ash (Raask, 1985,
Clarke and Sloss, 1992).

The major problems resulting from the ash content are concerned with deposition and corrosion (slagging and fouling) in the furnace. The fly ash has to be
removed from the flue gases since it is a major nuisance rather than a major pollution hazard, although there is an increasing interest in the emission of ultrafine particles, especially of toxic metals that do constitute a health hazard. Some
of the ash is in the form of cenospheres, but generally, ash is emitted as a powdery dust or larger grit particles. Apart from the fly ash the particles may form
deposits on metal surfaces, but most of the ash deposits at the bottom of the

TABLE 3.1
PRINCIPAL CHEMICALS FOUND IN COAL MINERAL MATTER

Chemical group	Properties
Clays minerals	Produce inert ash and form gases
Sulfides and sulfates	Produce corrosive sulfur compounds
Carbonates	Produce carbon dioxide
Silicates	Produce glasslike materials
Chlorides	Produce glasslike materials, also corrosive compounds
Metallic oxides (mostly iron)	Form inert materials
Trace elements (including rare metals)	Can form corrosive substances at high temperatures, also poisonous fumes

TABLE 3.2
COAL ASH ANALYSES FOR TYPICAL ASHES

Component (wt%)	High rank (bituminous)	Low rank (lignite)
SiO_2	40–55	5–15
Al_2O_3	18–25	3–15
Fe_2O_3	5–20	5–10
CaO	5–8	3–30
MgO	1–10	2–5
Na_2O	1–2	5–15
K_2O	1–2	0.1–0.5
TiO_2	1–2	0.1
P_2O_5	1–3	0.1
SO_3	1–3	20–30

combustion chamber or grate and is removed for disposal either in landfill or for construction purposes.

Table 3.3 outlines the sizes of particles that are directly emitted by two principal combustion methods, the chain grate being large particle and from pulverized coal combustion. Clearly, the particulates have to be collected from flue gases, and various particulate collection systems are used. The principal methods of collection and particle-size range collected are cyclones (5–200 µm), bag filter (0.01–10 µm), and electrostatic precipitation (0.01 to 50 µm). In present-day power stations the collection efficiencies are as high as 99.5%.

The mechanism formation of fly ash formation in pf combustion is outlined in Figure 3.4. The processes are complicated by melting (fusion), fragmentation, vaporization, and coalescence.

Mineral matter in coal also contains the toxic metals listed in Table 3.4. These

TABLE 3.3
PARTICULATE MATTER SIZE DISTRIBUTION IN FLUE GASES

Particle size (μm)	Chain grate (% wt)	Pulverized fuel (% wt)
0–5	1.0	17.8
5–10	1.5	13.5
10–15	2.0	11.9
15–20	2.0	9.9
20–30	4.5	14.3
30–40	5.0	9.4
Over 40	84.0	23.2
Totals	100.0	100.0
Typical particulate concentrations (g/m³) STP	0.5–4.5	10–14

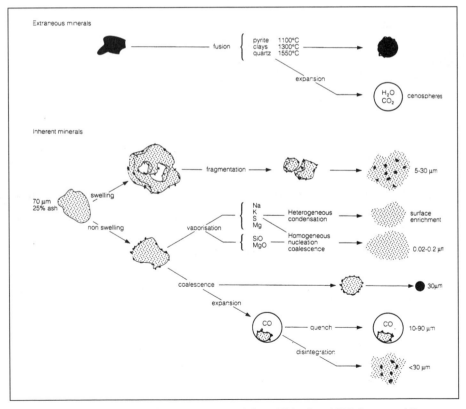

Figure 3.3   Mechanisms for fly ash formation (after Wibberley, 1985, Jones and Benson, 1998).

TABLE 3.4
OCCURRENCE OF TRACE ELEMENTS IN COAL

Element	Average concentration (order of magnitude, in ppm)	Main form of occurrence	Emission characteristics
Antimony	1	Sulfide	Enrichment
Arsenic	10	Sulfide	Enrichment
Beryllium	1	Organic	
Boron	100	Organic	Enrichment
Cadmium	1	Sulfide	Enrichment
Chlorine	1000	Organic and inorganic	Vapor
Chromium	10		Enrichment
Cobalt	10	Organic and inorganic	Refractory
Copper	10	Sulfide	Enrichment
Fluorine	100		Vapor
Lead	10	Sulfide	Enrichment
Manganese	100	Carbonate	Refractory
Mercury	0.1	Sulfide	Vapor
Molybdenum	1	Organic and inorganic	Enrichment
Nickel	10	Organic and inorganic	Enrichment
Selenium	1	Sulfide	Vapor
Thallium	1		Enrichment
Thorium	1		Refractory
Titanium	1000	Aluminosilicate	Refractory
Tungsten	1		
Uranium	1		Enrichment
Vanadium	10	Organic	Enrichment
Zinc	100	Sulfide	Enrichment

can be nonvolatile and emitted with the ash, or they can vaporize and be emitted as a gas (such as mercury) or they can evaporate and recondense in the cooler flue gases. In this latter case they become enriched as indicated in Table 3.4.

The electrostatic precipitator (ESP) is the most versatile collection device, and in it the electrodes charge the ash particles with a negative charge that then adhere to the negative collecting plates, where they are removed by mechanical agitation at specific intervals. Their efficiency is a function of the sulfur content of the coals and the fly ash resistivity (which should be low) and cohesivity, the latter two being determined by the amount of unburned carbon.

3.2
Formation of Carbon Monoxide

Carbon monoxide is found in the combustion products of all carbonaceous fuels, generally in concentrations well above that expected from equilibrium considerations. For any system that is in equilibrium the carbon monoxide concentration is given by the overall reaction

$$CO_2 \underset{}{\overset{1}{\rightleftarrows}} CO + \tfrac{1}{2}O_2 \quad K_{3.1}$$
R 3.1

and thus $[CO] = K_1[CO_2]/[O_2]^{1/2}$, where K_1 is the equilibrium constant and $[CO]$, etc., are the concentrations (in appropriate units) of the gases. Values of $K_{3.1}$ may be determined from JANAF tables, and some other values are given in Appendix 2.

It is clear that the equilibrium level of carbon monoxide is dependent on the temperature and the level of excess air. Low levels of excess air result in higher concentrations of carbon monoxide if the temperature is maintained constant.

Carbon monoxide is formed in flames by the rapid oxidation of hydrocarbons by oxygen in the reaction zone or by the oxidation of char in the postflame region. The carbon monoxide is subsequently slowly oxidized to carbon dioxide by reactions (R3.2) and (R3.3), which combined form the water-gas equilibrium reaction

$$CO + OH \underset{}{\overset{2}{\rightleftarrows}} CO_2 + H$$
R 3.2

$$H + H_2O \underset{}{\overset{3}{\rightleftarrows}} H_2 + OH$$
R 3.3

$$CO + H_2O \underset{}{\overset{4}{\rightleftarrows}} CO_2 + H_2$$
R 3.4

Since the carbon monoxide is formed rapidly in the reaction zone, but only slowly consumed, the concentrations of carbon monoxide present in the reaction zone are above the equilibrium values. The slow conversion of carbon monoxide to carbon dioxide in the postflame zone gases is termed afterburning. The rate of this conversion is given by

$$\frac{-d(CO)}{dt} = 4.0 \times 10^{14}[CO][O_2]^{1/4}[H_2O]^{1/2} \exp\left(\frac{-20,202}{T}\right) \text{mol/cm}^3/\text{s}$$

where $[CO]$ is the concentration of CO, etc. This expression can be integrated to give the concentration of CO at any time t during combustion if the initial concentration is known.

If the time available for burn-out of the carbon monoxide is short, as in small combustion chambers, the concentrations of carbon monoxide in the burned gases are higher than for large units. Thus, the CO levels for small units are about 1000 ppm compared with 50–100 ppm for large combustion chambers, e.g., of the type found in power stations.

3.3
Pollutants Originating from Sulfur Present in Coal (SO$_x$)

Formation of Sulfur Dioxide

All coals contain some organosulfur compounds that are present as sulfides, disulfides, or cyclic compounds together with inorganic compounds. Their nature and concentration is dependent upon the origin of the coal and vary between 0.1 and 10 wt% as discussed in Chapter 1. The emission index of SO$_2$, and the overall amount of ash, is readily calculated from the analysis of the fuel, and such a calculation is given in Appendix 3.

On combustion, the sulfur compounds are thermally unstable (S—S or S—C bonds readily break) and are rapidly converted to SO$_2$ in the oxidizing conditions in the flame zone (Figure 3.5). The extent of SO$_2$ formation (in ppm) is given approximately by $510 \times$ % sulfur in the coal for stoichiometric combustion so that typically, SO$_2$ concentrations in the stack gases are in the range of 200–2000 ppm. The sulfur dioxide produced is undesirable for a number of reasons, but the principal one is that it or products derived from it (SO$_3$, H$_2$SO$_4$ aerosol, and acid rain) are air pollutants that can damage health, cause corrosion, and cause acidification of lakes. Little can be done by combustion control methods to minimize the emission of SO$_2$. The only control strategies involve eliminating the sulfur from the fuel or removing the SO$_2$ from the combustion gases in situ by means of a sorbent or in the flue gases by flue gas desulfurization. Further SO$_2$ reacts to form SO$_3$, and minimization of its formation is very much a control matter. Since the removal of the SO$_2$ from the stack gases is a difficult and costly operation, its presence is the principal factor determining chimney heights for stationary installations. Details of methods of estimating chimney heights and the problems associated with dispersion are given in a number of accounts.

Formation of Sulfur Trioxide

Sulfur trioxide may be formed from the sulfur dioxide produced initially, the mixture of SO$_2$ and SO$_3$ being termed SO$_x$. Sulfur trioxide is undesirable in that it may react with water to form sulfuric acid. Since the acid dewpoint, or temperature at which condensation occurs, may be as high as 150°C, condensation may occur in economizers and air heaters, resulting in the corrosion of metal surfaces and the formation of deposits of sulfates. In addition, it causes a visible

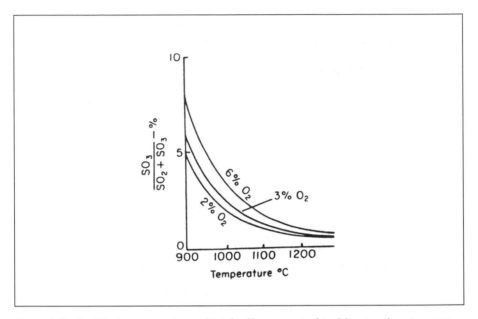

Figure 3.5 Equilibrium percentage of total sulfur converted to SO_3 at various temperatures and oxygen concentrations.

plume from the stacks due to the formation of enhanced water condensation or sulfuric acid aerosols. This problem can be overcome by minimizing the formation of sulfur trioxide or by the use of additives, while corrosion can be minimized by maintaining all metallic surfaces above the acid dewpoint.

The overall reaction for the oxidation of SO_2 is given by

$$SO_2 + \tfrac{1}{2}O_2 \rightleftharpoons SO_3 \quad \Delta H^{\circ}_{298} = -98.28 \text{ kJ/mol} \qquad\qquad \text{R 3.5}$$

but the actual mechanism is more complicated. It is now well established that two reactions are involved; the first involves the association reaction with an oxygen atom, namely,

$$SO_2 + O + M = SO_3 + M \quad E = 0 \text{ kJ/mol} \qquad\qquad \text{R 3.6}$$

where M is a third body, and the second is the reaction

$$SO_2 + OH = SO_3 + H \quad E = 0 \text{ kJ/mol} \qquad\qquad \text{R 3.7}$$

where reaction (R3.6) is dominant in flame zones and reaction (3.7) at lower temperatures. The reaction may be catalyzed by solid surfaces, particularly carbon, or by transition metals such as deposited Fe. Since these are usually present on surfaces, the SO_3 produced can remain as complex salts such as $Na_3Fe(SO_4)_3$, which can play an important role in corrosion and slagging. Estimates can be made of both homogeneous and heterogeneous SO_2.

The homogeneous reactions are moderately rapid under most flame situations and equilibrium is fairly rapidly established, so that the concentrations of SO_3 in flame or stack gases may readily be calculated from the equilibrium

$$SO_2 + O_2 = SO_3 + H \quad E = 300\,kJ/mol \qquad\qquad R\ 3.8$$

Hence
$$[SO_3] = K_{3.8}[SO_2][O_2]^{1/2}$$

In the temperature range 1000–2000K, K_8 is given approximately by log $K_{3.8} = 5014/T - 4.755$, where the concentrations are given in bars. The concentration of SO_2 can be calculated from the sulfur content of the fuel, and since the oxygen content in the burned gases is known by direct measurement or by calculation, the amount of SO_3 present can be deduced. It is clear that the concentration of SO_3 increases with the level of excess air (i.e., with $[O_2]^{1/2}$). Generally, the concentration of SO_3 in flue gases is about 0.2–3% of the total sulfur oxides present as shown in Figure 3.5, and thus SO_3 concentrations rarely exceed 50 ppm.

The Formation of Acid Smuts and Their Prevention

This problem is restricted to stationary combustion units and usually smaller boilers, in which carbonaceous residues formed as a result of combustion may be carried to the wall of a flue and are deposited on the wall at some part where the surface temperature is such that sulfuric acid from the combustion gases has already condensed on it. Deposits of acid-soaked carbonaceous materials then build up, but due to changes in temperatures or gas flow, small particles break off and are emitted with the gases from the chimney.

As the flue gases cool, SO_3 combines with water vapor to produce sulfuric acid vapor, and below about 600K almost all the SO_3 exists as H_2SO_4:

$$SO_3 + H^+ + OH^- = 2H^+ + SO_4^{2-} \qquad\qquad R\ 3.9$$

There is considerable variability both in the theoretical and the experimental data on the sulfuric acid dewpoints, T_{DP}. The recommended equation is

$$\frac{1}{T_{DP}} = 2.276 \times 10^{-3} - 2.943 \times 10^{-5} \ln P_{H_2O} - 8.58 \times 10^{-5} \ln P_{H_2SO_4}$$

$$= 6.2 \times 10^{-6} \left(\ln P_{H_2O} \right) \left(\ln P_{H_2SO_4} \right)$$

where T_{DP} is in kelvin and the partial pressures are in mmHg. Some typical sulfuric acid dewpoint curves for some specified flue gas compositions are shown in Figure 3.6.

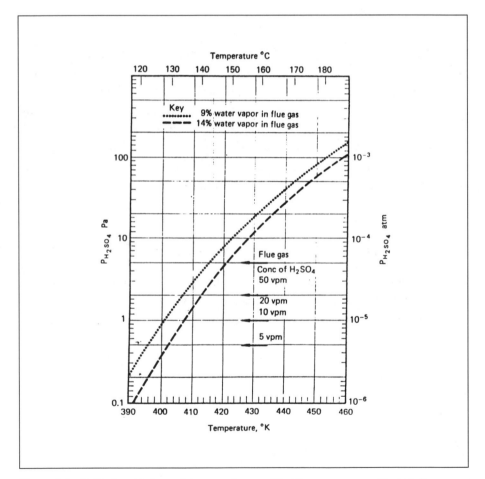

Figure 3.6   Sulfuric acid dewpoint curves for specified flue gas compositions (after Davies et al., 1981).

At the present time a common form of the additive is $Mg(OH)_2$ powder. This may be added to the fuel for deposit and corrosion control, but if it is added specifically to control SO_3 formation it may be added after the combustion chamber as a fine powder (5-mm diameter) and injected into ducts through pneumatic lines at a molar ratio for Mg/SO_3 of about 2/3. It reacts with the SO_3 produced to produce magnesium sulfite, and the magnesium also coats any surfaces that are capable of catalyzing the conversion of SO_2 to SO_3. The aluminum modifies the ash produced so that any deposits readily flake off (i.e., are friable), and it also reduces the formation of soot and carbon. There are no known adverse health effects arising from simple Mg-based additives, but the rate of additive injection is limited by the legal limit on chimney solids emissions. While there is an economic penalty in their use, this is usually acceptable.

Control of Sulfur Dioxide Emissions

Considerable attention has been devoted to the adverse effects of acidic rain and deposition generally, which is largely attributed to combustion-generated products, and as a consequence, legislation in many countries limits the emission of SO_2 at least from large combustion plant (the EU limit is for a 50-MW plants but may be reduced in the future).

The effect of sulfur can be minimized by

1. Reducing the formation of sulfuric acid by using fuels with a low sulfur content.
2. Minimizing the concentration of SO_3 by reducing the excess air as previously outlined.
3. Using additives. In this technique materials that react with SO_3 or H_2SO_4 are added. Powdered dolomite ($MgCO_3 \cdot CaCO_3$) or limestone ($CaCO_3$) is used as sorbent.
4. Using flue gas processing. This is called flue gas desulfurization (FGD).

Many processes have been investigated, but at the present time, processes based on using limestone as an absorbent seems to be the most effective for coal- (or oil-) fired plants (e.g., Babcock-Hitachi Process).

The SO_2 is simply removed by injecting a spray of limestone-water slurry into an absorber tower as indicated in Figure 3.7 (Babcock Hitachi Process).

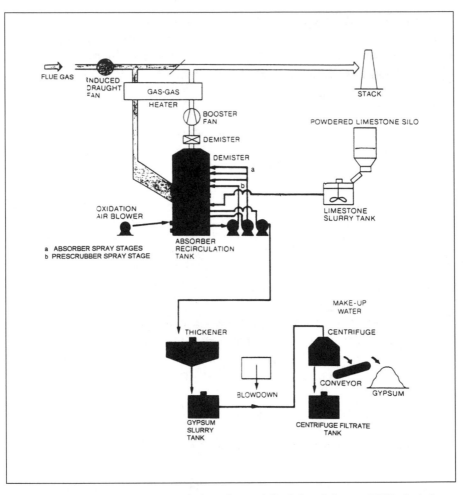

Figure 3.7   Diagrammatic representation of one of the Babcock Power DESO$_x$ Emission Control Systems.

The reactions are

$$SO_2 + H_2O \rightarrow H^+ + HSO_3^-$$ R 3.10

$$H^+ + HSO_3^- + \tfrac{1}{2}O_2 \rightarrow 2H^+SO_3^{2-}$$ R 3.11

$$CaCO_3 + 2H^+SO_4^{2-} + 2H_2O \rightarrow CaSO_4 \cdot 2H_2O(s) + CO_2$$ R 3.12

The oxidation process in the reaction uses excess oxygen present in the flue gas, and this supplemented by injecting air into the slurry in the sump at the base of the absorber. The efficiency of removal of SO_2 is 90–96% and the gypsum formed is devolatilized and is removed as dry crystalline gypsum ($CaSO_4 \cdot 2H_2O$) suitable for use in wallboard or cement manufacture. In China, some steps are being taken to use it to improve soil that has too high a sodium content.

HCl gas that is present in the flue gas is soluble in the reagent and reacts with the calcium carbonate, producing calcium chloride ($CaCl_2$). Because of its preferential reactivity, high levels of $CaCl_2$ would inhibit the SO_2 removal process and reduce the quality of the gypsum ($CaCl_2$ normally less than 100 ppm). The buildup of $CaCl_2$ is therefore controlled. The cleaned flue gas is then reheated with incoming flue gas to about 80°C before being discharged via the stack.

3.4
Formation and Control of Oxides of Nitrogen (NO_x)

Mechanism of Formation of NO_x

During combustion of fuels with air, a small part of the nitrogen present in the air or in the fuel itself reacts with oxygen to form nitric oxide in the flame gases. This nitric oxide reacts further in the cooling gases or when the combustion products leave the combustion unit to form NO_2 (and to a limited extent N_2O_4); the mixture of these oxides of nitrogen so formed is called NO_x. In addition, a small amount of N_2O is formed during the combustion processes mainly because some of the NO originally formed is reduced by carbon particles in the flame.

The formation of NO_x in flames involves three routes. The first is the well-established thermal route that is often termed the Zeldovich mechanism. The second involves the reaction of fuel hydrocarbon fragments with molecular nitrogen and is termed the prompt-NO route, and the third is the reaction of organic-nitrogen compounds present in the coal, and its contribution is dependent upon the origin of the coal. These steps are discussed next and outlined in Figure 3.8.

Thermal Route

In the thermal mechanism, oxygen atoms that are present in the flame zone and combustion products react thus:

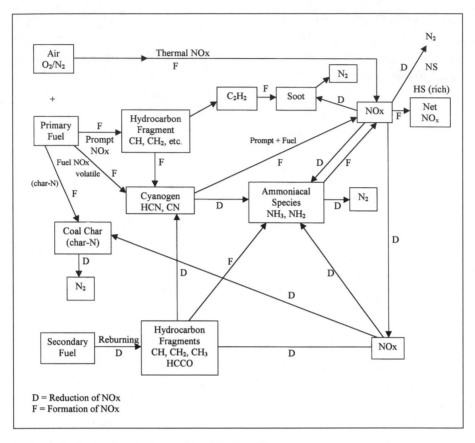

Figure 3.8   Routes leading to the formation of NO$_x$.

$$O + N_2 \rightleftharpoons NO + N \qquad\qquad\qquad \text{R 3.13}$$

$$N + O_2 \rightleftharpoons NO + O \qquad\qquad\qquad \text{R 3.14}$$

$$N_2 + O_2 \rightleftharpoons 2NO \; K_{3.15} \qquad\qquad\qquad \text{R 3.15}$$

Hence

$$[NO] = \sqrt{K_{3.15}[N_2][O_2]}$$

Values of $K_{3.15}$ may be derived from the JANAF tables.

The rate of reaction is such that it is rate-controlling; furthermore it is very temperature-dependent so that nitric oxide is formed only in high-temperature gases. Since the overall reaction is slow, equilibrium concentrations of NO are built up only in situations where there is a long residence time; that is, in large boilers. In smaller combustion units the nitric oxide concentration is limited by the lower residence time. The concentration of nitric oxide, if produced only by means of this thermal mechanism, can be calculated from the expression

$$\frac{d[NO]}{dt} = 2K_{3.15}[O][N_2] \quad -(\text{reverse reaction as equilibrium is approached})$$

$$= 1.4 \times 10^{17} T^{-1/2} \exp\left(\frac{-9460}{T}\right)[O_2]^{1/2}[N_2]$$

$$- 10^7 T^{-1/2} \exp\left(\frac{-46,900}{T}\right)[NO]^2[O_2]^{-1/2} \quad \text{mol/cm}^3/\text{s}$$

This expression can be integrated to give the concentrations of nitric oxide produced after time t_0. The time required for the equilibrium levels of NO to be established is such that the actual concentrations attained are only one-third to one-tenth of the equilibrium concentrations. This of course varies depending upon the circumstances, differing markedly between stationary plant applications operating at atmospheric pressure and engines that operate at high pressures. In practical situations the nitric oxide formed is most radically influenced by operating temperature, but it is also influenced by the level of excess air and the residence times. In boilers, the NO_x produced is approximately proportional to the firing rate, and so in practice the NO_x in combustion gases from coal-fired equipment ranges from 200 ppm for small installations to 1000 ppm for larger units. The upper figure is determined to some extent by the amount of fuel-N compounds present as described in the next section.

Prompt-NO

A considerable quantity of the NO_x formed in combustion is produced by the flames surrounding individual particles. There is a dependence of NO emission upon the particle diameter, but the overall factors are complex. In particular it has been shown that in pf combustion the finer coal particles produce less nitric oxide than coarser particles.

The reason for this is that generally much of the combustion occurs under fuel-rich conditions, i.e., surrounding burning devolatilizing particles. In these circumstances a certain amount of nitric oxide is produced by the so-called prompt-NO route. Here, carbon-containing free radicals react with molecular nitrogen to form nitric oxide by reactions that probably involve the following:

$$
\begin{array}{llllll}
\text{CH} & + \text{N}_2 & = & \text{HCN} & + \text{N} & \text{R 3.16} \\
\text{N} & + \text{O}_2 & = & \text{NO} & + \text{O} & \text{R 3.17} \\
\text{HCN} & + \text{OH} . & = \text{`} & \text{CN} & + \text{H}_2\text{O} & \text{R 3.18} \\
& & & \text{HNCO} & + \text{H} & \\
\text{CN} & + \text{O}_2 & = & \text{CO} & + \text{NO} & \text{R 3.19} \\
\text{O} & + \text{HCN} & = & \text{NCO} & + \text{H} & \text{R 3.20} \\
& & = & \text{CO} & + \text{NH} & \\
& & = & \text{OH} & + \text{CN} & \\
\text{HNCO} & + \text{H} & = & \text{NH}_2 & + \text{CO} & \text{R 3.21} \\
\text{NO} & + \text{NH}_2 & = & \text{N}_2 & + \text{H}_2\text{O} & \\
& & & & & \text{R 3.22}
\end{array}
$$

Consequently, coal flames contain HCN and NH_3 from subsequent reactions of NH.

Evidence for this comes from the fact that hydrogen cyanide is formed during the combustion of coal particles under fuel-rich conditions and is also present in the flame surrounding burning single particles. The conversion route for prompt-NO in a summarized form is shown in Figure 3.8. The cyanocompounds formed are converted either to NO or to CO and N_2.

Fuel-N

All coals contain cyclic organic-nitrogen compounds analogous to indoles and pyridines that are termed fuel-nitrogen (fuel-N) compounds (Burchill and Welch, 1989). The distribution of the fuel-N between the volatiles and char, which is discussed in the next chapter, is an important factor. Under pyrolysis conditions these compounds form the volatile components and decompose to give HCN (and a little CH_3CN) together with a char-containing nitrogen (char-nitrogen):

Coal $\rightarrow$ volatiles (including HCN, CH_3CN) + char (including N) R 3.23

Nitric oxide may also be produced indirectly from the fuel-N compounds but in small amounts, so that in combustion nitric oxide is produced by two routes, the direct formation and the thermal (Zeldovich) route.

The HCN then reacts via reaction R3.18 onward to give NO or N_2. Not all the fuel-N is converted to NO; some produces nitrogen. The relative yields of NO and N_2 depend upon the concentration of the fuel-bound nitrogen in the

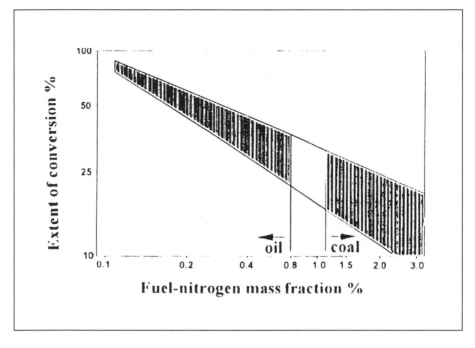

Figure 3.9 Extent of conversion to NO as a function of fuel-N content of fuel oils and coal.

fuel and the stoichiometry in the volatile combustion zone; if rich, all the HCN is converted to N_2.

The char-N forms NO and N_2 in an analogous way, the ratio being determined by the stoichiometry and the kinetics. Even so, the presence of fuel-N in coals may increase the concentration of nitric oxide in the stack gases by many hundred parts per million. The influence of the fuel-N on the contribution to the total NO_x production in a boiler is shown in Figure 3.9. These levels are also dependent on the stoichiometry of the combustion system and on the reaction time.

Details of the chemical mechanism of the conversion of fuel-N to nitric oxide are now fairly well established, and it is known that the process is less temperature-dependent than the thermal route and that reduction in flame temperature does not markedly reduce nitric oxide formed in this way.

The key factors in NO_x formation can be summarized in Figures 3.10 and 3.11. Figure 3.10 shows the contribution of the different routes to NO_x formation, and fuel-N dominates.

Figure 3.11 shows the levels of NO formation in flue gases as a function of

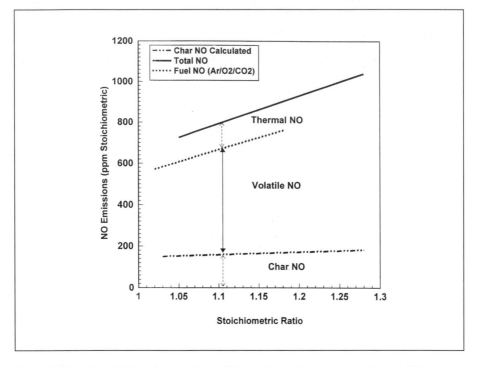

Figure 3.10 Plot of NO emissions form different formation routes against stoichiometric ratio (after Smart et al., 1998).

the fuel-N, clearly closely related. Figure 3.11 shows how the extent of conversion to NO decreases as the fuel-N mass fraction increases, but since the fuel-N of most coals lies in a small band, then this factor does not change too much. However, some coals may have a low fuel-N content (e.g., Indonesian) and some high (e.g., some Australian coals).

Control of NO_x Formation by Combustion Modification

As previously outlined, in recent years very considerable attention has been directed to the control of NO, NO_2, and N_2O emissions. The primary pollutant NO is converted to NO_2 after emission by atmospheric oxygen (and ozone). NO_2 can be involved in acid rain (up to 30%) after conversion to HNO_3. It is also involved in the production of photochemical oxidants via the reactions

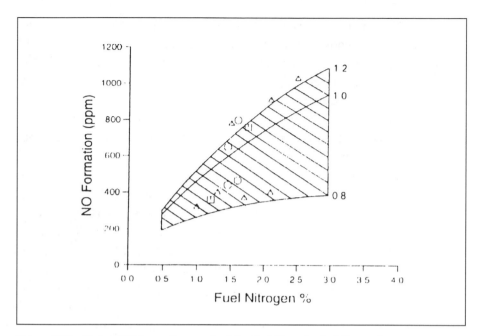

Figure 3.11   Plot of NO formation against the fuel-N content of the coal.

$$NO_2 + h\mu = NO + O \qquad\qquad R\ 3.23$$
$$O + O_2 + M = O_3 + M \qquad\qquad R\ 3.24$$
$$O_3 + \text{hydrocarbons} \rightarrow \text{organic photochemical oxidants} \qquad R\ 3.25$$

The ozone and other photochemical oxidants produced via NO_2 have adverse effects, such as damaging trees and producing photochemical smog, and as a result legislation has been introduced to limit the emission of NO. In boilers, the NO_x produced ranges from 100 ppm for small installations to 1000 ppm for larger units such as thermal power stations as previously described. The upper figure is determined to a major extent by the amount of fuel-N compounds present in the coal, and most large installations, power stations for example, operate with levels in the 800-ppm range unless low-NO_x burners are installed. The general trend via legislation is to bring the emission levels down to about 100 ppm. N_2O is also of concern because it is implicated as a "greenhouse" gas causing atmospheric warming, but more importantly, N_2O can react with atomic oxygen and cause damage to the ozone layer; unlike NO_2 it is not rained out at lower altitudes. The amount of N_2O emitted varies with the source, from about 10 ppm in pf combustion to several hundred ppm in fluidized-bed combustion.

As a consequence of the detrimental effects of these emissions, a number of control measures have been instigated, some of which involve the nature of the combustion process and others that involve flue gas or exhaust gas treatment.

Since oxygen atom concentrations and flame temperatures are low in fuel-rich stoichiometric situations, a number of NO_x control methods have attempted to adopt this technique. Generally, this involves two-stage (staged) combustion; the first stage of combustion is undertaken under rich conditions and in the second stage additional air is added to complete combustion. In this way the oxygen atom concentration as well as the peak temperature is reduced, and thus the rate of NO formation is reduced. However, a major difficulty is that there is a tendency for soot to be formed and considerable care has to be exercised in its application. In continuous-flow combustion chambers such as furnaces and boilers, this method is readily applied by operating one or more burners rich and adding additional air later on. This is also a form of staged combustion. Alternatively, in multiburner arrays some burners are run rich and other burners are used to provide the additional air. A variation of two-stage combustion is "reburn" in which the last stage is run rich either by coal particle injection or natural gas injection. The rich fuel components (CH, CH_3, etc.) react with the NO reducing it to N_2 via R3.16, namely,

$$CH,\ CH_3 + NO = HCN + H,\ H_2O \qquad\qquad \text{R 3.25a}$$

$$HC\ \ CO + NO = CN + CO_2 \quad \text{at low temperature} \qquad \text{R 3.25b}$$

An alternative approach is to use external exhaust gas recirculation in which some of the exhaust gases are recycled, and this technique is applicable to stationary units. Here, the recirculation of oxygen-deficient combustion products results in some reduction in the flame temperatures and there is a consequential reduction in NO formation. In addition, some of the NO present in the recirculated gases is destroyed on its passage through the flame zone by the reaction. Furthermore, soot formation may be reduced as previously described. However, the additional equipment necessary for the gas recirculation makes its application difficult or unsatisfactory for some applications.

Flue Gas Treatment

A number of processes have been developed to decompose NO in the flue gases, and these are outlined with the above techniques in Table 3.5.

TABLE 3.5
TYPICAL COST OF VARIOUS NO$_x$ CONTROL TECHNIQUES (1998)

	Cost in $ per kW of thermal input		
Technology	Installation	Operation/year	NO$_x$ reduction
Combustion modification			
1. Low-NO$_x$ burners	3	Low	30–50
2. Furnace air staging	3.6	Low	Up to 50
3. Combination of low-NO$_x$ and air staging	5.7	Low	Up to 70
4. Reburn gas or coal	9–21	1.5	Up to 60
Flue gas treatment			
5. Selective noncatalytic treatment (NH$_3$ injection)	3.3–16	2.2–3	40–50
6. Selective catalytic treatment (NH$_3$ injection with catalyst)	21–27	4.5 and over	80–90
7. NO$_x$, SO$_x$			

Injection of Ammonia and Related Compounds

Thermal deNO$_x$ selective noncatalytic reduction (SNCR) is the most well known of the available processes in which ammonia is injected into flue gases. The reactions are controlled by the processes

$$NH_3 + OH = NH_2 \qquad\qquad R\ 3.26$$
$$NH_2 + NO \rightarrow N_2 + H_2O \qquad\qquad R\ 3.27$$

The kinetics and reaction mechanism are generally well known. Because of the necessity of maintaining an adequate supply of NH$_2$ radicals, the process has a temperature window of about 900–1100°C, limited at the upper end because the NH$_2$ is oxidized to NO and limited at the lower temperature because the reaction becomes too slow. This thermal window can be widened or lowered by various additives. An interesting aspect is that N$_2$O is not emitted by this process.

The process is increasingly used, but the main practical difficulty is the injection of ammonia that has to be undertaken in such a way that it mixes uniformly with the flue gases; this is necessary to maximise the amount of NO reacted and to prevent the emission of unreacted NH$_3$. The use of wall-jet injectors has been successful in this respect. Usually, the NH$_3$ to NO ratio is 1.5/1.

Because of the need for high temperatures in the thermal deNO$_x$ process, a lower temperature catalytic process has been developed, known as the selective catalytic reduction method (SCR). Here, ammonia is injected into the flue gas stream at about 500°C and reaction takes place in contact with a catalyst, typically iron oxide with some other transitional metal oxides present as accelerators. Poisoning of the catalysts can be a problem, and much depends on the nature of the fuel being used.

Isocyanic Acid (RAPRENO$_x$) and Urea

In the RAPRENO$_x$ process, solid cyanuric acid is vaporized to HNCO (isocyanuric acid), which reduces NO selectively at temperatures as low as 450°C. Basically the mechanism involves the following steps:

$$HNCO + H = NH_2 + CO \qquad\qquad \text{R 3.28}$$
$$NH_2 + NO, \text{ etc., as for thermal deNO}_x \qquad \text{R 3.29}$$
$$HNCO + OH = NCO + H_2O$$

followed by NCO reactions that produce NH$_i$ radicals with subsequent NO destruction. Urea can also be used, and the NO$_x$OUT process based on this has been used on gas-fired plants and in principal could be used for coal plants. In this, the first step is the decomposition of the urea; thus:

$$(NH_2)_2 CO = HCNO + NH_3 \qquad\qquad \text{R 3.30}$$

That is, urea acts as a source of isocyanic acid and ammonia and they react by the mechanisms outlined. In practice urea is injected as a powder into the hot flue gases, where the urea evaporates and decomposes giving the reactants. However, in terms of cost the ammonia must be the preferred additive.

Interaction of SO$_x$ on NO$_x$ Formation

The influence of fuel-sulfur on nitrogen oxide formation has been previously investigated. Although a number of different mechanisms have been proposed for sulfur interaction by various workers, the main consensus is that the presence of sulfur results in sulfur compounds such as SO, SH, H$_2$S, S$_2$, CS$_2$, and COS existing in superequilibrium in postflame regions. It has been suggested that sulfur presence accelerates recombination of H radicals and those of O and OH atoms, thus resulting in reduced thermal-NO formation by the extended Zeldovich mechanism. The effect of fuel-NO formation is more complex and less well understood, although it is believed to be strongly dependent upon flame stoichiometry and flame temperature. Future developments to the NO_x prediction model will incorporate the influence of sulfur on overall NO_x formation by assuming that in fuel-rich zones, sulfur intermediates will react according to

$$SO_2 + H \rightarrow HS, H_2S \qquad \text{R 3.31}$$
$$HS + NO \rightarrow OH + NS \qquad \text{R 3.32}$$
$$NS \rightarrow N_2 + S_2 \qquad \text{R 3.33}$$

Although the incorporation of these reactions into the NO_x package will result in NO_x reductions, the magnitude of these reductions is assumed to be small.

Reburn as an NO_x Control Strategy

Reburn is an in-furnace or flue gas NO_x control technique whereby a small amount of a hydrocarbon fuel, such as natural gas or coal, is injected into the products from the main (primary zone) combustion zone as illustrated in Figure 3.12. The amount of primary to secondary fuel dictates the staging ratio. The secondary zone forms CH and CH_3 radicals from the fuel that are responsible for the reduction of NO to N_2. A simplified reaction mechanism is of the form

$$CH + NO = HCN + O$$
$$\text{R 3.16}$$
and $$NO + CH_2 \rightarrow HCN + H_2O$$

HCN formed may then react to form N_2 via the reactions set out earlier.

The efficiency of the reburn technique is determined by the initial stoichiometry of the primary zone. If the resulting oxygen concentrations are too high, then HCN formed in the reburn process may proceed to form NO. Furthermore, the injection of burn-out air after the initial reburn zone may result in NO_x formation via

$$HCN \xrightarrow{\text{OH,O}} NO + CO \qquad \text{R 3.34}$$
$$NH_3 \xrightarrow{\text{OH,O}} NO + OH \qquad \text{R 3.35}$$
$$NH_3 \xrightarrow{\text{OH,O}} NO + O \qquad \text{R 3.36}$$

The mechanisms that govern the formation and destruction of NO during re-

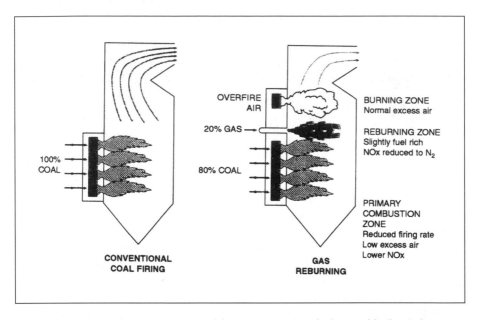

Figure 3.12   Operating parameters and furnace zones in a fuel-staged boiler (reburn).

burning were based on studies by Wendt and Pershing (1977) and Chen at al.(1982).

The following expression can be used to describe NO decay:

$$\frac{d[NO]}{dt} = -[NO][NH_3]\, f_1 - [NO][CH_4]\, f_2$$

where functions f_1 are functions of temperature and OH and H_2O concentrations, i.e.,

$$f_1 = \sum_j A_{ij} T^{N_{ij}} \exp\left(\frac{B_{ij}}{T}\right)\left(\frac{[OH]^{m_{ij}}}{[H_2O]^{l_{ij}}}\right)$$

The rate of change of NH_3 and HCN are obtained from the following expressions:

$$\frac{d\,[HCN]}{dt} = -[HCN][\, f_3 + f_4] + [CH_4]\{[NO]f_2 + [N_2]f_5 + [NH_3]f_6\}$$

$$\frac{d\,[NH_3]}{dt} = -\frac{d\,[NO]}{dt} - \frac{d[HCN]}{dt} - 2\frac{d\,[N_2]}{dt}$$

Industrial Application

The preceding discussion has been largely general in nature, but its application to a particular combustion plant depends upon the exact circumstances. Computational modeling techniques are now widely applied to these problems, especially for NO_x minimization in the combustion chamber, for reburn, and for flue gas treatments. The development of low-NO_x coal combustion systems will be examined in detail later with particular emphasis on the use of computational modeling techniques to aid combustor design. In order that predictions of pollutant formation in a combustion chamber be accurate, the basic coal combustion program must be correct, i.e., accurate temperature, oxygen concentration, and char burnout predictions throughout the furnace. Herein lies a fundamental problem in that the availability of in-flame coal combustion data is scarce and difficult to obtain. As such, alternative methods of data acquisition for coal combustion are used based on computer submodels as discussed in Chapter 4. The representation of the processes of devolatilization and char burn-out as commonly used in computational packages is highlighted and future developments can include effects such as fly-ash formation, deposition, slagging, etc., and also trace metals. NO_x predictions from coal combustion systems are of particular importance. The accurate prediction of prompt-, thermal-, and fuel-NO are necessary in order to have a useful design/operating tool that can aid in minimizing pollutant formation from combustion plant. Fuel-NO_x formation is recognized as being the major component of total NO_x emissions from such processes. With this in mind, inaccuracies in predicting overall emissions are most dependent upon the fuel-NO_x predictive routines. The same can be said for all the other pollutants, but NO_x has attracted considerable attention because NO_x can be markedly reduced by in-chamber combustion modification.

The rapid development of computer capabilities will result in expanded computer utilization as both a design and operating tool. As processing times decrease, the incorporation of more complex and representative coal combustion routines into programs such as CFD codes becomes feasible. The main problem arising in coal combustion modeling is, however, the heterogeneous nature of the coal. Nevertheless, computer programs are now available that predict the whole operation of coal-fired plants ranging from the economics of operation, the influence of blending, and the effect on pollutants.

4

COMBUSTION MECHANISM OF PULVERIZED COAL

4.1
Role of Pulverized Fuel Combustion

At the present time, most of the world's coal is used for electricity production, and this operation mainly involves the combustion of pulverized coal. Indeed, the importance of this type of combustion overall seems to be leading to a worldwide increase in coal use. Clearly, pollutant emissions from coal combustion for power generation will be a significant component of overall fossil fuel emissions for a considerable time, and consequently, extensive research has been undertaken on pf combustion much more than for other forms of coal combustion.

During the 1980s the control of the emissions of SO_x and NO_x attracted considerable attention, and increasingly stringent controls have been applied since then. The global emissions of NO_x are about 20 Mt and, if the projected increase in coal consumption occurs, will increase by about 10% by the year 2020 unless extensive control measures are adopted. A similar situation holds for SO_2 emissions. A number of techniques are available for NO_x and SO_x control, and these have already been highlighted in the previous chapter and in Table 3.6, NO_x emission attracting considerable attention because it is in principal a problem soluble by combustion modification. This arises from the fact that the treated NO is emitted as molecular nitrogen, a component of air. The treatment of SO_2 is more difficult because the products are sulfur, sulfuric acid, or gypsum. A use has to be found for the last product as a building material, or it has to be landfilled; the former two materials can be used in the chemical industry. There is in particular a major need for an innovative use for sulfur. The relative costs and reduction efficiencies for each process are major factors in determining which method is used as a pollutant-controlling process.

In the case of NO_x reduction, it is believed that the necessary reduction measures can be obtained by a combination of in-furnace combustion modification, low-NO_x burners, reburn, etc., without the need for expensive flue gas treatment processes. Combustion modified low-NO_x systems have the additional benefits of being easy to maintain once installed and do not require the handling of toxic chemicals, such as ammonia, associated with flue gas NO_x treatment techniques; furthermore, it should not result in an increase in unburned carbon. This chapter therefore concentrates on the development of the understanding of pf combustion from the point of view of flame performance and the combustion chamber modification techniques necessary for NO_x control strategy and to enable the high levels of carbon burn-out. In addition, consideration has to be directed to flame stability, minimization of CO emissions, the influence of the flame on boiler types in terms of lifetime (erosion and corrosion), the level of excess air that determines smoke levels and heat loss in the stack gases as well as a number of other minor pollutants such as VOCs, PAHs, dioxins, and toxic metals. It sets out the bases for the use of computational codes as a design method to predict combustor performance and emissions from coal combustion systems currently in use and under development. These applications will be discussed in Chapter 6.

Extensive work has been undertaken over the last 20 years or so concerned with studies of coal-particle devolatilization, char properties, and char burn-out rates. Prerequisites for pollution formation in coal combustion processes involve understanding factors governing smoke formation, unburned carbon, and ash aerosols as well as the conventional pollutants SO_x and NO_x.

A number of computational codes have been developed and utilized with regard to coal combustion modeling. These models must account for heat transfer effects to the coal particle, coal-particle heat-up, devolatilization, and subsequent reaction of the volatiles to CO, char reactions, and ash formation. Coupling of these with fluid flow calculations in order to represent the complex aerodynamic behavior present in most burner systems must be made prior to pollutant formation modeling. NO_x formation predictions from coal processes using computational programs have been reported previously and require the incorporation of thermal- , prompt- , and fuel-NO formation mechanisms. Nitrogen contained within the coal must be modeled during its release as both volatile and char-bound nitrogen and the ensuing reactions represented. A similar situation applies for sulfur and toxic metals.

The successful development of computational codes is dependent upon the acquisition of accurate experimental data for comparison, especially when utilizing extensive CFD packages to predict the combustion characteristics and pollutant formation within utility systems. Currently, most predictive methods produce qualitative results that are sufficient to examine trends in burner operation, for example, although the need for quantitative predictions still remains.

The main processes of coal combustion, namely, coal devolatilization and char burn-out, are usually simplified to the following reactions:

Step 1: $Coal = char + volatiles$ R 4.1

Step 2: $Volatiles (HC) + O_2 \rightarrow CO + H_2O$ R 4.2

Step 3: $CO + \frac{1}{2}O_2 = CO_2$ R 4.3

Step 4: $\phi C(char) + O_2 \rightarrow 2(\phi - 1) CO + (2 - \phi) CO_2$ R 4.4

More advanced combustion models are now available and these are discussed in the following sections.

Their development and incorporation into combustion prediction computational codes are discussed together with future possible modifications under development. Ultimately one would like a code that can input coal properties (such as ultimate analysis, structural parameters) and compute through to combustion chamber performance, pollutant formation, and include an estimate of unburned carbon. A schematic diagram of the steps required in a detailed model of coal combustion is shown in Figure 4.1.

It should be stressed that although considerable attention is directed here to the combustion mechanism, the important adjuncts of heat transfer and flame stability are vital for successful combustion.

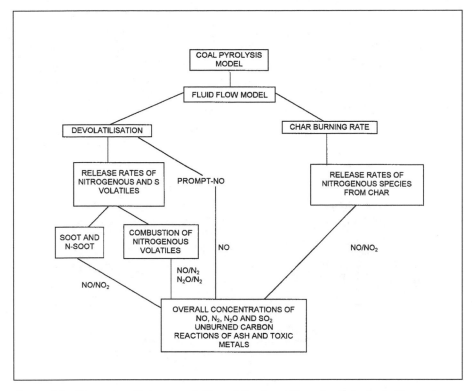

Figure 4.1 Diagrammatic representation of the formation of pollutants during coal-particle combustion.

4.2
Devolatilization of Coal Particles

A key combustion step in coal combustion is the rapid heating of coal particles and their devolatilization in the near burner zone. The flow patterns are complex and usually involve swirling flow, recirculation, and high radiant fluxes, while the topological features of the softening coal particles change, influencing particle drag and trajectories. The heating process is rapid as shown for a typical case in Figure 4.2. Various publications describe heating rates as about 10^5 K/s and this seems reasonable as an average type of figure; but clearly, particle size plays an important role. The coal particle undergoes decomposition into char and volatile material, the former burning slowly in the later stages of the flame, while the volatile material, at its simplest, is assumed to rapidly form CO and subsequently CO_2. This process of devolatilization involves the break-

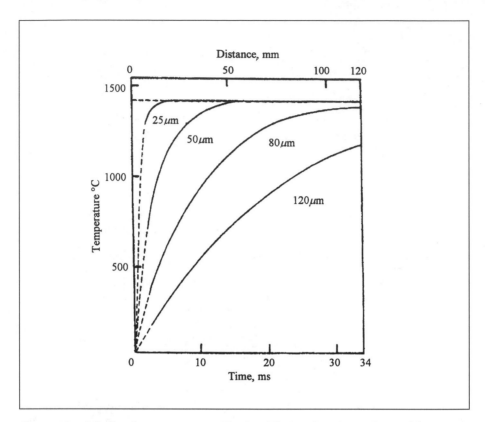

Figure 4.2   Calculated temperature profiles for different diameter coal particles in pyrolysis flame at 1675K (after Unsworth et al., 1991).

down of the initial coal particle into a complex mixture of light gases, tars, etc., that are ejected and subsequently react to form not only CO and CO_2 but also soot. Figures 4.3(a) and (b) show two stages of a devolatilizing particle. Their behavior markedly depends on their fluid properties when starting to react. Measurements of swelling behavior are important; indeed, some very plastic molten coal particles can disintegrate during combustion, causing disruptive devolatilization and disruptive combustion.

Experimental Studies

There are a number of theoretical approaches, but these naturally require experimental data to set up and/or validate them. Such data have primarily been

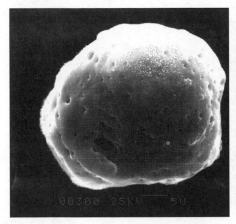 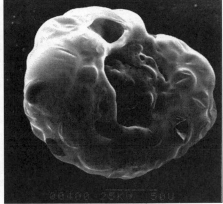

Figure 4.3(a)   Coal particle undergoing pyrolysis in the early stages of a flame.

Figure 4.3(b)   Coal particle undergoing pyrolysis in the latter stages of a flame.

obtained from studies of single suspended coal particles, a heated wire mesh, or drop tube experiments. It is possible to obtain kinetic data on devolatilization, as illustrated in Figure 4.4, but there are differences depending on the method used. While these systems have certain advantages, they can be problematical in their operation; i.e., before the volatiles can be analyzed, they have to be collected, which in the case of heavier tars may result in condenzation upon cool surfaces and the need for removal requiring special solvents. Extensive measurements have been made on tar yields and on the chars produced. Such experiments provide useful data for new or unusual coal that present problems, e.g., Colombian coal, part of that burns out slowly.

An alternative experimental technique for products, one providing a fingerprint of the coal, involves the use of a coal pyrolysis unit coupled to a gas chromatograph (pyrolysis/GC). Typically, this would consist of a Pyroprobe or Curie point pyrolysis instrument that uses a small platinum heated coil that can have controlled heating rates and hold times. A coal sample of about 3 mg is heated so that any evolved gases can pass directly into the gas chromatograph column, thus eliminating collection problems. The apparatus can be operated at heating rates of up to 2.10^K/s, up to temperatures of 1500°C, which are comparable with the temperatures attained by wire meshes (Azhakesan et al., 1991).

The coal pyrolysis products are extensive and produce four distinct groups of volatiles that may be classified according to molecular mass, namely: gases and light, medium, and heavy tars. This means that coals themselves can be classi-

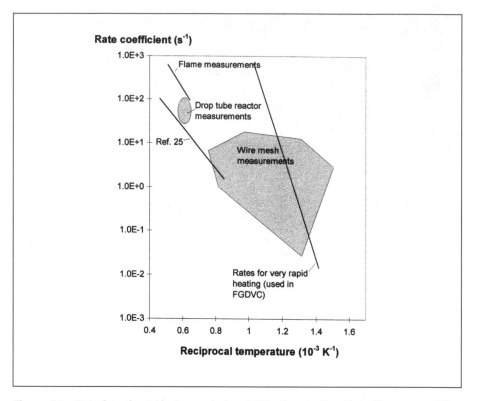

Figure 4.4   Data from heated wire mesh devolatilization studies. Plot of log rate coeffi-cient versus reciprocal temperature.

fied according to their relative yields of each fraction. Figure 4.5 shows a typical analysis of some of the lighter species. It is interesting to note that the nitrogen is released mainly as HCN. The composition of the volatiles changes with coal rank, gas yields steadily decrease with increasing rank, and all three tar yields reach a peak at around 70% carbon content (daf), which corresponds to the high volatile bituminous coals.

The yield of the heavy tars from rapidly heated coal particles can be studied by heated wire mesh experiments or by FIMS (field ionization mass spectrometry). A typical FIMS analysis of the tars evolved is given in Figure 4.6. The composition of the tars can be obtained much more simply from the difference between the coal composition and the char composition (which can be assumed to be almost 100% carbon).

An alternative method of studying devolatilization is by means of an entrained flow or drop tube reactor. In these, coal particles are entrained in an air

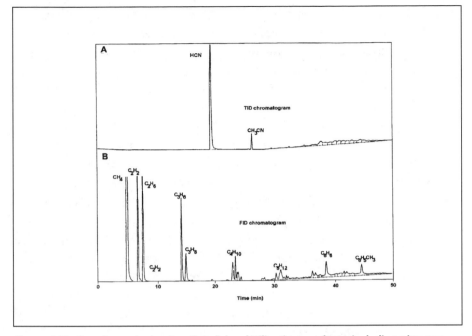

Figure 4.5   Gas chromatography of coal devolatilization products, including nitrogen products (A) and hydrocarbons (B)

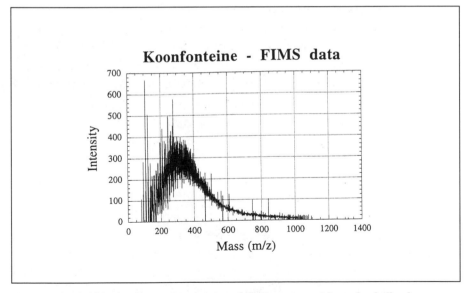

Figure 4.6   Field ionization mass spectrum of the tars present in coal volatiles from Koonfonteine (S. Africa) coal (by permission of AFR).

stream and passed along an electrically heated reactor tube. Samples are removed and analyzed and the components C, H, N, etc., are determined as a
function of the ash content. At the temperatures normally used by drop tubes,
1000–1500°C, it is assumed that there is no vaporization loss from the ash.
Typical curves showing the variation of the major species are shown in Figure
4.7. Figure 4.7(a) shows the changes occurring under oxidative conditions with
a high heating rate in a drop tube reaction and involves sampling along the axis
of the reactor. The H content drops only slowly, and one of the problems is the
fact that coal and coal char are not well-defined separate entities, and that there
is a slow convergence to the "char," which in the first instance is "young" char.
In a pf flame this distinction is still unclear since the flame formed from the
emitted volatiles will collapse on to the surface, and oxidation of the surface
will start even before the particle has turned to char. Measurements have also
been made under pyrolytic conditions and these are shown in Figure 4.7(b). It is
interesting to note the increase in the nitrogen in the char in this case (Clarke et
al., 1987) as the reaction progresses. This concentration effect of the residual nitrogen occurs both at high temperatures, as in the case here, and at fluidized-
bed temperatures where the effect is well established.

Coal Devolatilization Computational Models

In order to model this process, two main theoretical approaches are adopted.
The first involves the construction of a detailed model that accounts for the decomposition of the coal matrix during coal-particle heat up. This approach requires detailed analytical data to be obtained from the specific coal to be
modeled, describing both the initial coal matrix structure and the release of the
gases and tars during pyrolysis. This approach has been taken by a number of
researchers and is described by the work of Solomon et al. (1993), Solomon and
Fletcher (1994), Smith et al., (1994), Niksa and coworkers: e.g., Niksa (1993,
1996), and Beer (1996).

The second approach involves modeling the devolatilization process using
generalized expressions relying on a reduced set of chemical reactions, and commonly the two-step mechanism is used. This approach is more commonly
adopted within CFD codes that would become restrictively slow if large coal
matrix pyrolysis programs were utilized in the main body of the computation.
The data required in order to represent the devolatilization process must therefore be carefully selected so as to be both easily computed and readily obtain-
able, while maintaining the ability to account for coal heterogeneity.

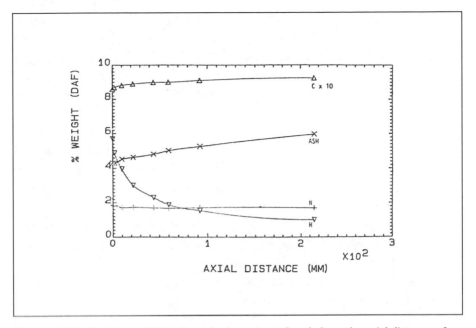

Figure 4.7(a)   Variation of C, H, N, and ash content of coal along the axial distance of a CH$_4$/air flame (high heating rate)

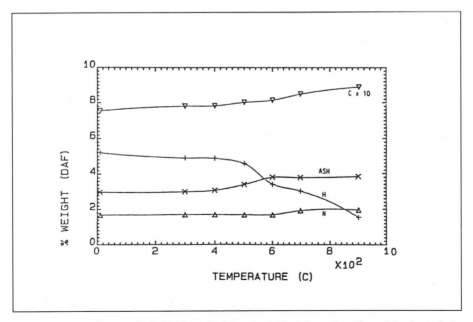

Figure 4.7(b)   Changes in C, H, N, and ash content of coal as a function of final pyrolysis temperature (low heating rate)

Ideally, by inputting coal properties it would be useful to predict, by means of a preprocessing computer subroutine, the coal devolatilization behavior. One such model is the FG-DVC (functional group-depolymerization vaporization cross-linking) computational model (Solomon, 1979, Solomon and Fletcher, 1994); other similar models are FLASHCHAIN (Niksa, 1996) and CPD (Smith et al., 1994).

The basis of the FG-DVC computer model is a set of coals from the US Argonne Premium Coal Samples Bank, although this has been extended recently to internationally traded coals. FG-DVC is a two-stage program, designed to predict the rates and yields of all the major gas species, and the yields and elemental compositions of the tars and char from a coal undergoing pyrolysis. It is accurate at heating rates of up to 40 K/ms and is capable of modeling very slow heating rates, as would be experienced by coal under storage. Indeed, it could model coal from source through to its utilization.

The functional group part of the program is based on the premise that a fraction of the total gases evolved is produced when functional groups break away from the macromolecular network. Other functional groups stay attached to the main coal structure, eventually being released as part of a light tar molecule. The fraction of functional groups released as gases are determined by the heating rate, final temperature, functional group density, and structural properties of the original coal. A typical reaction scheme for this process that involves the breaking of cross-linking and ring groups in a coal molecule of the type given in Chapter 2. Small groups would produce gases, larger molecular fragments would give tars.

The second stage of the program considers thermal breakdown of macromolecular networks. It starts with an approximation of a network, including all the major structural attributes; molecular weights of the aromatic ring clusters, cross-link density, and the potential number of labile bridges. The decomposition of a simple two-dimensional network is undertaken according to a Monte Carlo simulation, and the network starts to fragment. The smaller fragments form tars and gases that evaporate. What remains of the macromolecule forms the char.

FG-DVC uses as its basis libraries, each describing the chemical, structural, and kinetic properties of one coal. FLASHCHAIN uses a similar system, although it is based more on the physical steps, namely, the distillation of the tars and other products. Herein lies the greatest limitation of current models since, although it can predict with considerable accuracy the yields and composition of volatiles from a coal undergoing pyrolysis, it is limited to the coals within its database or an interpolation of their properties. The information contained within each library is highly involved and complex, and to create a new library for a different coal involves substantial effort and research into the properties of that coal. Thus the behavior of "unusual" or unknown coals is difficult to undertake at present until universal generalized equations are available.

The computer models, FG-DVC, FLASHCHAIN, and CPD have proved to be worthwhile tools and have given results consistent from a drop tube or wire mesh experiments. An example of the comparison between experiment and computed values is given in Figure 4.8. Figure 4.8(a) gives the percentage yields under the stated conditions and Figure 4.8(b) shows all the distribution of the nitrogen. The effect of temperature and heating sites on product yields and distribution of nitrogen—or any other species—can readily be obtained by FG-DVC or FLASHCHAIN. An example of such a computation is shown in Figure 4.9(a) to (d).

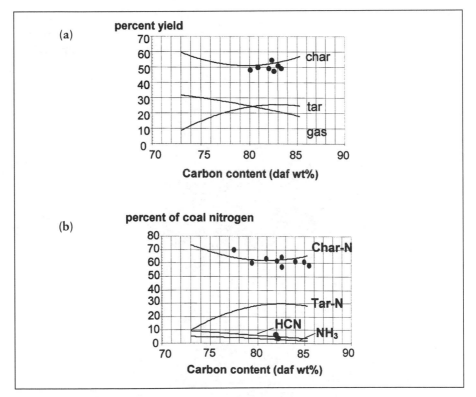

Figure 4.8 Predicted (lines) and observed (dots) products from devolatilization of coals of varying rank (10^4 K/s to 1273K). (a) Char, tar, and gas yields, (b) char-N, tar-N, HCN, and NH_3.

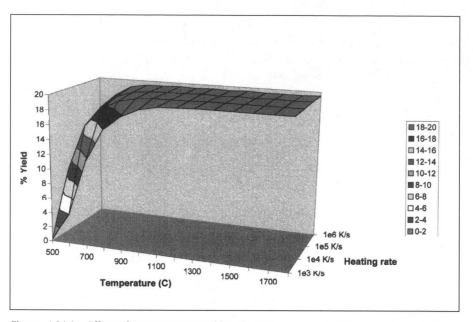

Figure 4.9(a)   Effect of temperature and heating rate on gas yield from coal pyrolysis.

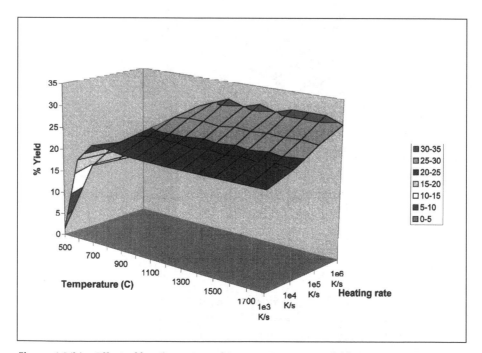

Figure 4.9(b)   Effect of heating rate and temperature on tar yield.

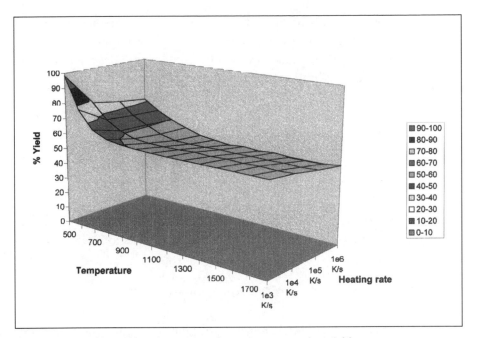

Figure 4.9(c)   Effect of heating rate and temperature on char yield.

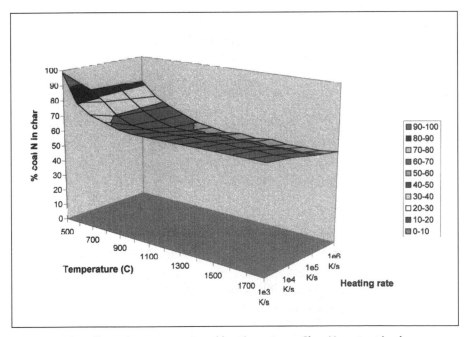

Figure 4.9(d)   Effect of temperature and heating rate on Char-N content in char.

Volatile release-rate representation While a preprocessor computation of the amount of volatiles and their composition is of vital importance, in order to model the rate of release of volatiles within CFD codes a more simplistic representation of the devolatilization process is required. A number of different expressions have been developed, with most relying on Arrhenius rate parameters to determine the rate of volatile release. The Badzioch and Hawksley model (1970) relies on a single Arrhenius expression to represent volatile release according to

$$\frac{dm_p}{dt} = m_{p0}\left(V^* - V\right)A_0 \exp\left(\frac{-E_0}{RT_p}\right) \qquad\qquad \text{E 4.1}$$

Alternative models utilizing two or more Arrhenius expressions are frequently used within CFD codes, i.e., the two competing reaction mechanism developed by Kobayashi et al. (1976):

$$\frac{dm_p}{dt} = m_{v0} f_v \left(\alpha_2 R_1 + \alpha_2 R_2\right)\exp\left[-(R_1 + R_2)t\right] \qquad \text{E 4.2}$$

R_1 and R_2 are competing rates that may control the devolatilization over different temperature ranges:

$$R_1 = A_1 \exp\left(\frac{E_1}{RT_p}\right) \qquad\qquad\qquad \text{E 4.3}$$

$$R_2 = A_2 \exp\left(\frac{E_2}{RT_p}\right) \qquad\qquad\qquad \text{E 4.4}$$

This type of model relies on six parameters to describe the devolatilization of the coal and should be adaptable to modeling most coals, providing the data are available. It has certain benefits over the Badzioch and Hawksley model in that the total volatile release is a function of the particle temperature. Earlier data from wire mesh experiments and from drop tube experiments suggest that pyrolysis is represented by relatively low activation energies. The present evidence indicates that these data have been influenced by the effects of heating rates and that high activation energies are appropriate. Indeed, for many models it is possible to assume that the mass lost is simply a function of the rise in temperature (Weber, 1996). In the case of the FG-DVC model, it is possible to estimate the rates from the time-dependent curves of volatiles emissions.

4.3
Combustion of Volatiles

The nature of most models is such that the evolving volatile species are represented by a single compound that may lead to inaccuracies, and speciated models are better. Future developments of devolatilization routines thus require a more detailed approach in which additional volatile materials such as tars may be represented, and that the gases are speciated into at least groups of compounds. One approach is to use measured or computed gaseous volatiles as input into a flamelet model and use the SANDIA OPP DIFF computer programs (US SANDIA Laboratories) to calculate the reaction. These data can be used to construct a library that can be accessed by the main computer program. Figure 4.10 illustrates such a computation. The tar yields can be assumed to be converted to soot, which then burns independently outside the flamelet. The soot composition can be estimated using the assumption that hydrogen is lost from the tar by the process of soot aging.

4.4
Char Burn-out

The accurate modeling of char burn-out must account for a number of different factors such as char temperature, local oxygen concentrations, residence times within the reaction zone, and time/temperature histories of each particle modeled. Additional considerations relating to char reactivity must be made, which may include coal type and petrography, surface area, char porosity, particle-size distribution, and ash blockage/catalytic effects. Coals form chars of different types, some of which are shown in Figure 4.11. Some char combustion models require an input of the surface area of the char. Sometimes this is assumed to be 300 m^2/g, but in general, this is a function of rank and can be represented by the expression $(0.368C^2 - 54.4C + 2109)$ m^2/g where C is the percentage wt of the initial coal (daf) if it assumed that CO_2 surface areas are appropriate. If N_2-BET surface areas are used, their values are much lower.

The combustion process initially produces CO and CO_2, this ratio being markedly dependent on the temperature. More CO is produced at high temperatures, and at higher temperatures the direct formation of CO_2 at the surface becomes negligible. The ratio of CO to CO_2 is given by $[CO]/[CO_2] = 2500$

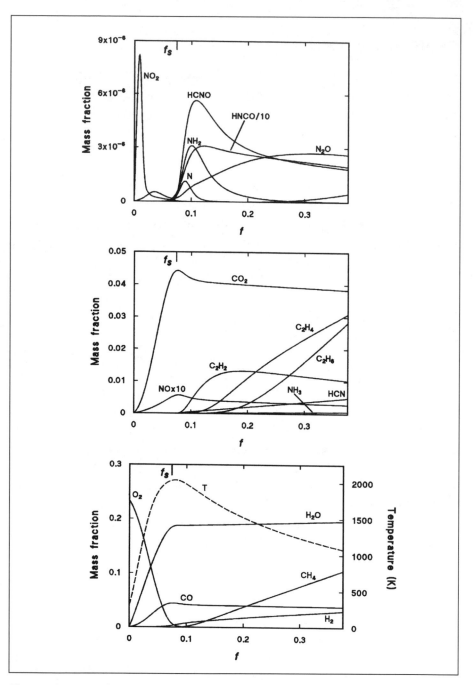

Figure 4.10 Mass fraction and temperature profiles as a function of mixture fraction (f) around a burning coal particle.

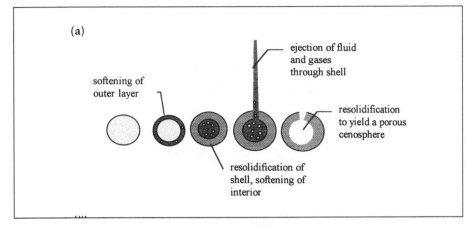

Figure 4.11(a) **Stages in the formation of a porous cenosphere.**

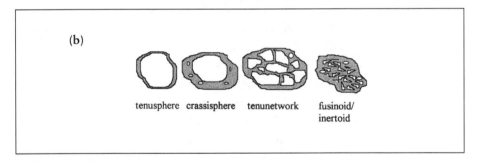

Figure 4.11(b) **Types of cenosphere produced.**

$\exp(-6262.6/T_p)$, where the ratio has the value of about 12 at 900°C according to the original study by Arthur (1951). Subsequent work on this ratio and the mechanism has been compiled by Zeng and Fu (1996); based on this and additional work, $[CO]/[CO_2] = 600 \ (\rho_s X_{O_2}, s)^{0.24} \exp (-8000/T_p)$, where $X_{O_2}s$ is the mass fraction of oxygen on the surface and (as above) $[CO]/[CO_2]$ is the molar ratio; X_{O_2}, s can be approximated to the bulk oxygen concentration but can be calculated explicitly; ρ_g is the gas density (kg/m^3).

The carbon oxidation reaction mechanism is still not completely elucidated in terms of the elementary processes of adsorption, complex formation, rearrangement, and desorption. The complexity of individual steps is compounded by the structural diversity of the carbon skeleton complicated by the presence of N, S, and mineral matter. Various mechanisms have been proposed with various degrees of sophistication, but all involve the formation of surface oxygen complexes.

For the combustion temperatures considered here, the production of CO and CO_2 during carbon combustion can be simplified in the following scheme:

R4.5

$$C_b + \tfrac{1}{2}O_2 \xrightarrow{k_1} \tfrac{1}{2}(C-O_2)_m \xrightarrow{k_2} C(O)_m \xrightarrow[k_3]{+C_e} C(O)_1 (+C_b) \xrightarrow{k_4} CO (+C_e)$$

$$k_5 \downarrow \; + C(O)_m \text{ or } C(O)_1$$

$$CO_2 (+2C_e \text{ or } C_e \text{ and } C_b)$$

where k_1 = rate of adsorption of molecular O_2; k_2 = rate of dissociation of adsorbed O_2 to mobile surface oxygen species $C(O)_m$; k_3 = rate of surface diffusion of $C(O)_m$ to a localized active site C_e [reaction to $C(O)_1$] assumed to be instantaneous; k_4 = rate of desorption of CO from localized sites $C(O)_1$; and k_5 = rate of surface reaction of two neighboring $C(O)_1$, or one $C(O)_m$ and one $C(O)_1$ and desorption of CO_2.

The first steps are adsorption of molecular oxygen onto free basal sites of the carbon C_b, and dissociation to produce "mobile" surface oxygen species $C(O)_m$. These species can diffuse over the defect-free part of the carbon surface toward the edges of the aromatic regions, or defect sites, C_e. Here they transform into localized surface oxides $C(O)_1$, which can either desorb as CO or combine with neighboring localized or mobile oxygen to yield CO_2, this determining the ratio of CO/CO_2. In both cases, free active sites C_e are regenerated on the surface. For the present model, it is assumed that desorption of CO_2 is faster than desorption of CO, due to thermodynamic considerations. Similarly, the reversible adsorption of molecular oxygen k_1 is assumed to be rapid. This being the case, then if (k_1 to k_3) $>>k_4$, the localized and mobile surface atomic oxygen species concentration Θ_O is high, and more CO_2 is produced relative to CO. In contrast, if (k_1 to k_3)$<< k_4$, then Θ_O is low, and more CO is produced relative to CO_2.

The effect of heat treatment on the mechanism of combustion is a result of the effect of annealing and ordering the carbon lamellae. According to the model, this increased order in the carbon structure will increase the reservoir of molecular and possibly atomic oxygen species, and may also influence the rate of surface diffusion k_3. Thus, Θ_O is higher and the CO/CO_2 ratio decreases.

The evolution of NO during the oxidation of the nitrogen-containing carbons is often described by active site theory as follows:

$$C_{fas} + C(N) + O_2 \rightarrow C(NO) + C(O) \qquad \text{R 4.6}$$

$$C(NO) + C(O) \rightarrow NO(g) + C(O) + C_{fas} \qquad \text{R 4.7}$$

Where C_{fas}, $C(N)$, $C(NO)$, and $C(O)$ are a free active site of the carbon, surface nitrogen, surface nitrogen oxide, and surface oxide complexes on the carbon. The nitric oxide produced during oxidation of fuel-N can be reduced on the carbon as it diffuses through the porous structure (Aarna and Suuberg, 1997, 1998). This involves the free active carbon sites and oxygen surface species set out below:

$$NO + 2C_{fas} = C(N) + [CO \text{ or } C(O)] \qquad \text{R 4.8}$$

$$2C(N) \rightarrow N_2 + 2C_{fas} \qquad \text{R 4.9}$$

$$C(O) + NO + 2C_{fas} = C(N) + [CO_2 \text{ or } C(O_2)] \qquad \text{R 4.10}$$

It is clear that whichever the mechanism, the rate of NO reduction could be dependent upon the steady-state concentration of free active sites, i.e., NO reduction could be sensitive to the surface oxygen coverage Θ_O. This would be the case if adsorption of NO onto the carbon were the rate-controlling factor.

Three different temperature zones of char combustion are generally considered as shown in Figure 4.12. In low-temperature environments (zone I), it is assumed that the oxidant diffuses readily into the char pores and that overall char reaction is governed by chemical kinetics. In medium-temperature regions (zone II), both chemical kinetics and oxidant diffusion play an important role (the chemical reaction rate is large enough to assume that the level of oxidant within the pores decreases to a value of zero at the particle center. In high-temperature regions (zone III), the kinetic rates are sufficiently high to assume that all the oxidant (oxygen) reacts on the char surface, and therefore the oxidant concentration decreases to 300 within the particle. The char reaction rate will thus be governed by the rate of oxidant diffusion from the bulk phase to the particle surface. This also has effect on the activation energy of the process as shown in Figure 4.12.

Generally, high oxygen partial pressures (and high pressures) tend to lead to chemical control, as do small particle sizes. The diffusional reaction rate coefficient, i.e., for the estimation of the degree of external diffusion control can be calculated as set out by Field et al. (1967) and by others (Smith, 1982) subsequently; it is also discussed later in this section.

The rate of reaction in the chemical-controlled region is known as the reactivity, defined by $-(1/W_0)(dW/dt)$, where W_0 = initial weight of carbon and

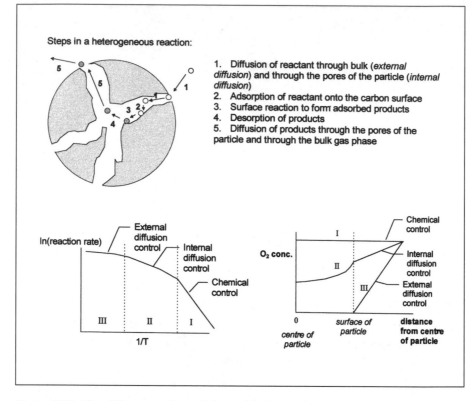

Figure 4.12   The different regimes of char oxidation: I. Chemical control, $E_{APP} = E_{TRUE}$, effectiveness factor $\eta = 1$, $M_{APP} = M_{TRUE}$; II. internal (pore) diffusion control, $E_{APP} = \frac{1}{2}E_{TRUE}$. $\eta < \frac{1}{2}$, $M_{APP} = \frac{1}{2}(M_{TRUE} + 1)$, III. surface diffusion control, $E_{APP} \rightarrow 0$, $\eta << 1$.

W = weight of carbon at time t. For coals, W_0 takes the value of the fixed carbon content.

The different nature of coals, and hence the variations in reactivity of the resultant char, makes the task of combustion modeling a complex one. Unfortunately, the required data for a variety of char particle sizes in large combustion systems may not always be readily attainable. In order to represent char burnout, char models are generally of a relatively simple nature and require data that is, on the whole, easily obtained. However, char combustion models often encompass a number of combustion regimes, taking place usually in zone II where diffusion and kinetic control come into play. However, the combustion zone is dominated by diffusion, whereas as the gases cool as they pass through heat exchanges, kinetic factor comes increasingly into play. The char burn-out model that has often been used for studies for coal combustion in pf flames assumes a model based on the original work by Field et al. (1967) and Baum and Street (1971). It is assumed that char burn-out commences once all the volatiles have evolved, but in reality there is overlap.

The mass burning rate can be represented by

$$\text{Char burning rate} = f \{A_t, \rho_c, C_O^n, k_i\}$$

where A_t is the reactive area and will depend on the mode (kinetic/diffusion control) of combustion, ρ_c is the density of the char (actually the density of the surface reactive carbon), C_O is the active oxygen concentration at the char surface, and k_i is the effective reaction rate coefficient. Various methods have been developed to express this rate and this has involved a variety of terms to describe reactivity, etc. In particular, confusion can occur because of the use of the term intrinsic reaction rate and of overall apparent rates.

The most widely used overall burning expression is that by Baum and Street (1971). The particle mass loss may be calculated from

$$\frac{dm_p}{dt} = -\pi D_p^2 \rho \left(\frac{RT_g X_{O_2}^n}{M_{O_2}} \right) R_T \qquad \text{E 4.5}$$

Here, the effective reaction area is the external volume of the char particle and hence an apparent reaction rate is used; since chars have varying complex cenosphere structures, the assumption of external area contains simplifying terms that ultimately are embodied in the kinetic values used. R_T is an overall reaction rate incorporating both kinetic and diffusive terms:

$$R_T = \frac{1}{(1/R_{\text{diff}}) + (1/R_c)} \qquad \text{E 4.6}$$

where R_{diff} is the diffusional rate coefficient and R_c is the chemical reaction rate coefficient per unit external surface area, being represented by

$$R_C = A_f \varphi \exp\left(-\frac{E_a}{R_T} \right) \qquad \text{E 4.7}$$

where φ is the ratio of reacting surface with external (equivalent sphere) surface area of the particle and

$$R_{\text{diff}} = \frac{2\varphi D_p M_c}{RT_g D_p} \left(\frac{T_p + T_g}{2} \right)^{0.75}$$ E 4.8

where φ is the mechanism factor of value 2 or 1, depending on whether the product is CO or CO_2.

Similar equations would apply to the combustion of soot. Particle size is of paramount importance. Thus a 200-μm-diameter particle at 1000°C and 2 mol% O_2 is 85% diffusionally controlled, with a 20-μm diameter it is marginally controlled by diffusion (60%), while at 2-μm diameter the combustion is 80% chemically controlled. Thus the burn-out of soot, and of the remnants of char (the carbon in the ash), are controlled by chemical kinetics.

There are a number of difficult and unresolved issues. The first is the order of the reaction with oxygen, n. This is a poorly understood concept and has a value that changes with temperature, being about 1.0 at high temperatures and 0.5 at normal combustion temperatures—and it has been argued that it is a fundamentally flawed concept (Essenhigh and Mescher, 1996). The second problem concerns the uncertainty in which zone combustion is occurring, and if it changes as particles become smaller; this has been described above. The third is the role of ash, which is described further below.

The char combustion model assumes that the reacting surface, whether internal or external, is of the same nature. The residue after char burn-out is assumed to be ash, which is subject to cooling as it leaves the combustion system. The model does not incorporate ash effects during the char burn-out process and assumes that the residual ash coverage is small. Although ash effects have received some attention within the modeling field, these have generally concentrated on the resistance to diffusion as opposed to the catalytic properties that may be in evidence. Modified rate expressions have been formulated to incorporate ash diffusional resistances and are of the type

$$R_T = \frac{1}{\dfrac{1}{R_{\text{diff}}} + \dfrac{1}{R_c}\beta^2 + \dfrac{1}{R_{\text{ash}}}\alpha}$$ E 4.9

where $$\beta = \frac{\gamma_p}{R_p} \quad \text{and} \quad \alpha = \left(\frac{\beta - 1}{\beta} \right)$$

The above example is sometimes termed the unreacted core shrinking model.

The value for ash resistance may be determined assuming a correlation proposed by Ishida and Wen (1968), namely, $R_{ash} = R_{diff} E^{2.5}$.

The activation energy data, etc., in the above formulation are "apparent," with $E_a = \frac{1}{2} E_{actual}$. It can be cast in a more fundamental way, namely,

$$\rho_c = R_c (P_{O_2})^n = R_i A_g \gamma 6_p \eta (P_{O_2})^n \qquad \text{E 4.10}$$

where R_i is the intrinsic reaction rate, A_g is the total (reactive) surface area of the char, α is a shape factor, and η is the effectiveness factor calculated using Thiele's modulus in a unimodal pore system. This is described in more detail in Hargrave et al. (1986). Generally, $E_{actual} \approx 160$ to 180 kJ/mol and $A \approx 50$ to 300 g/cm/s.

More general representations of char reactivity (R_c) have been developed in which laboratory measurements of char reactivity are correlated against physical and chemical properties of the char. Hampartsoumian et al. (1989) have reported char reactivity expressions in which 24 chars and cokes have been correlated to give an expression of the form

$$\ln R_c = -51.8 - 9.996 \ln \sigma_a + 9.484 A_g^{-1} + 7.38 \ln C + 0.939 H + 0.00761 T_g \qquad \text{E 4.11}$$

A modification to this expression was also highlighted that allows coals of different maceral contents to be represented. This requires the addition of a correction factor f_{mac} to the RHS of Equation E4.11:

$$f_{mac} = \left[1.4 (\text{Vit}_M + 0.83 \text{Vit}_{PS})\right] - \left[0.6 (\text{In}_R + 1.6 \text{In}_{IR})\right] \qquad \text{E 4.12}$$

which is related to the maceral content of the coal.

Expressions of this type offer a simplistic method of evaluating changes in burn-out times with variations in char type, although their accuracy is dependent upon the initial suite of chars used. A more sophisticated model has been developed by Charpeney et al. (1992). The expression they have used is

$$k = 4.74 \times 10^{-4} \exp (1049 \, H_{char}) \exp (0.28_{coal}) \, T \exp \left(-\frac{3200}{RT}\right)$$

Mitchell and Hurt (1992) have also proposed that $E = -5.9 + 0.355$ (wt% coal C) and that

$$\ln (R_{1750K}) = 2.8 - 0.076 \text{ (wt\% coal C)}$$

A number of other predictions are also available and have been reviewed by Carpenter (1998). These are based on automated petrographic measurements using reflectance and can be incorporated into a Reactive Index that can be used for coal combustion predictor models (e.g., CQIM, EPRI, NO_x-LOI predictor). Generally, it can be said that South African, Australian, UK, and most US coals have high reactivity. Some US and some South African coals have medium reactivity, and the low reactivity coals are mostly South African.

4.5
Development of Carbon Burn-out Models for High Levels of Carbon Burn-out

Although models utilized in current studies produce reasonable results, there is a requirement to incorporate more advanced models into the codes that will include ash diffusion effects and high levels of carbon burn-out. In addition, the effect of ash shedding from the char-particle surface is also under development. This may occur as a result of a combination of gravitational and rotational forces since the char particles may be subject to high-frequency particle rotation of the order of 1000 revolutions per second. The resultant forces acting on the char particle are determined from

$$F_g = m_{ash}g \qquad\qquad \text{E 4.13}$$

$$F_r = m_{ash}r_{char}\omega^2 \qquad\qquad \text{E 4.14}$$

An attractive force also acts due to surface tension, which may be represented by the (modified Young's) equation

$$F_a = \gamma(1 + \cos\theta)r_{ash}\sin\theta \qquad\qquad \text{E 4.15}$$

The mass of a surface ash particle may be calculated from

$$m_{ash} = \tfrac{1}{3}\pi\rho r_{ash}\left\{4 - \left[1 - \cos(\pi - \theta)\right]\left[\sin^2(\pi - \theta) + 1 - \cos(\pi - \theta)\right]\right\} \qquad \text{E 4.16}$$

The propensity for ash to shed from the char particle during combustion can thus be defined as

$$R_{sh} = \frac{F_r F_g}{F_a} \qquad\qquad \text{E 4.17}$$

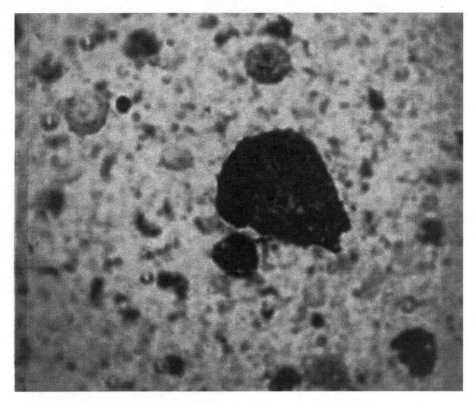

Figure 4.13 Unburned carbon in ash.

such that if

$R_{sh} > 1$, then ash will shed from the char particle

$R_{sh} < 1$, ash will be held on the particle surface

At high levels of burn-out, difficult problems occur with the earlier models because their accuracy of kinetic data doesn't permit extrapolation to 98% carbon burn-out. Because of the low levels of residual carbon, high precision is required to compute unburned carbon. At present this can be described by a probability approach and the use of a gamma-type function. The reactivity distribution function can be described using a gamma distribution proposed by Hurt et al. (1996) and Beeley et al. (1996):

$$F_A(A) = \left[\frac{e^{\beta A}}{\Gamma(A)}\right] A^{\alpha-1} \beta^{\alpha}$$

where α and β are distribution parameters, and $\Gamma(A)$ is a gamma function.

The variation in reactivity is represented by a variable A, and E and n are kept constant:

$$F(A, \rho, d_p, F, \text{In}) = F_A(A)F_p(\rho)F_d(dp)F_{Fr}(Fr)F_{In}(\text{In})$$

where F_{Fr} = distribution of particle fragmentation and F_{In} = distribution of ash effect. This involves a number of probability functions that must be determined experimentally.

An example of unburned carbon in ash is shown in Figure 4.13.

Nitrogen Oxide (NO$_x$) Pollutant Formation

There are a number of routes to NO_x formation from coal combustion that have been outlined previously. For a computational model to accurately predict the NO_x emissions from a given plant, all the major chemical reactions and physical properties of the coal must be included. The chemical composition of coals varies greatly. In the prediction of pollutant emissions, the nitrogen, sulfur, ash, and volatile matter contents are of particular relevance. Structural properties such as aromatic and aliphatic content and functional group concentration must also be considered. Once these parameters have been determined, the combustion chemistry can be analyzed. This can be broken down into two areas: homogeneous (volatile) and heterogeneous (char) combustion. The postprocessing NO_x package currently used in coal combustion predictions is subdivided into three main sections representing NO_x formation by thermal-, prompt-, and fuel-NO pathways.

Thermal-NO

Thermal-NO is modeled via the "extended Zeldovich" mechanism (Miller and Bowman, 1989) described earlier, namely,

$$N_2 + O \underset{k_{-11}}{\overset{k_{11}}{\rightleftharpoons}} NO + N \qquad \qquad \text{R 4.11}$$

$$N + O_2 \underset{k_{-2}}{\overset{k_2}{\rightleftharpoons}} NO + O \qquad \qquad \text{R 4.12}$$

but for fuel-rich mixtures, i.e., when OH > H > O, the additional reaction

$$N + OH \underset{k_{-13}}{\overset{k_{13}}{\rightleftharpoons}} NO + H \qquad \qquad \text{R 4.13}$$

must be added.

The rate-determining step governing the formation of practically all NO in the postflame gases is believed to be reaction R4.1. The rate constant used for this reaction is

$$k_{11} = 1.84 \times 10^{14} \ \exp\left(\frac{-38,370}{T}\right) \ cm^3/mol \ s$$

Providing there is a substantial amount of oxygen, i.e., a fuel-lean flame, the rate of consumption of free nitrogen will be equivalent to its rate of formation. Hence, making the "quasi-steady-state" assumption for the nitrogen atom concentration (Missaghi et al., 1991), one obtains

$$\frac{d[NO]_T}{dt} = 2k_{11}[O][N_2]\left(\frac{1-[NO]^2/K[O_2][N_2]}{1+k_{-1}[NO]/k_2[O_2]+k_3[OH]}\right) \qquad E\ 4.19$$

where $K = (k_{11}/k_{-11})(k_{12}/k_{-12})$ is the equilibrium constant for the reaction between N_2 and O_2.

Thermal-NO formation is highly dependent on temperature, linearly dependent on the oxygen atom concentration in the reaction zone, and associated with long residence times. Superequilibrium formation of O and OH atoms may occur in the flame front. If partial equilibrium is assumed, then one must account for the reaction

$$O_2 + M \underset{k_{-14}}{\overset{k_{14}}{\rightleftarrows}} O + O + M \qquad\qquad R\ 4.14$$

leading to the expression for oxygen concentration as

$$[O] = \lambda [K_O[O_2]]^{1/2} \qquad\qquad E\ 4.20$$

where λ determines the deviation from equilibrium and $k_{20} = k_{14}/k_{-14}$. It is necessary to incorporate this effect due to the strong influence O and OH atoms have in reactions R4.1 – R4.3.

However, for cases where thermal NO is the major contributor to the total NO (i.e., thermal-NO contribution to the total NO is higher than 30%), the O and OH concentration are based on the assumption of partial equilibrium of the chain-branching and propagation reactions of the H_2/O_2 mechanism, i.e.,

$$[O] = K_O \frac{[O_2][H_2]}{[H_2O]} \qquad\qquad E\ 4.21$$

$$[OH] = \{K_{OH}[O_2][H_2]\}^{1/2} \qquad\qquad E\ 4.22$$

where K_O and K_{OH} are the ratios of forward and reverse reaction rates of the $O + H_2$ reaction (see Appendix 2).

Prompt-NO

Prompt-NO is formed by reactions of N_2 with fuel-derived radicals in regions near the flame zone of hydrocarbon fuels. This type of NO occurs in fuel-rich, low-temperature environments and is associated with short residence times. Various species resulting from fuel fragmentation, for example, CH, CH_2, C_2, C_2H, C, etc., have been proposed as the precursor for prompt-NO formation. The major contributors are thought to be CH and CH_2:

$$CH + N_2 \rightleftharpoons HCN + N \qquad\qquad\qquad \text{R 4.15}$$
$$CH_2 + N_2 \rightleftharpoons HCN + NH \qquad\qquad\qquad \text{R 4.16}$$

The products of the above reaction could lead to the formation of amines and cyanocompounds, which may then react further to form NO:

$$HCN + O_2 \rightarrow NO + \ldots \qquad\qquad\qquad \text{R 4.17}$$
$$HCN_2 + N \rightarrow N_2 + \ldots \qquad\qquad\qquad \text{R 4.18}$$

The amount of prompt-NO is proportional to the number of carbon atoms present in the molecule of hydrocarbon fuel. Because of the complexity of the processes involved, a global kinetic parameter employing a reduced mechanism scheme is used to predict the prompt-NO emissions:

$$\frac{d[NO]_p}{dt} = fT^\beta k_{pr}[O_2]^a[N_2][\text{fuel}]^b \exp\left(-\frac{E_a}{RT}\right) \qquad\qquad \text{E 4.23}$$

where f is a correction factor for the fuel type and air/fuel ratio that is calculated from the expression

$$f = 4.75 + C_1 n - C_2 q + C_3 q^2 - C_4 q^3 \qquad\qquad \text{E 4.24}$$

where n = number of carbon atoms in the fuel, C_1, C_2, C_3, C_4 = 8.19×10^{-2}, 23.2, 32, and 12.2, respectively, and represents the non-Arrhenius behavior of the equation, q us the equivalence ratio, and β, a, and b are constants. Equation E4.24 has not been tested for high molecular weight species of the type present in coal tars.

Fuel-NO

In the prediction of NO_x from coal, nitrogen-bearing volatiles have to be considered separately, since these offer an extra route to NO_x formation: fuel-NO_x. As the nitrogen content of the coal increases, the amount of NO_x produced also rises. This can be seen in Figure 3.11, which shows predicted fuel-N conversions compared to experimentally obtained results. The fraction of total fuel-N converted to NO_x, however, decreases, and proportionally more molecular nitrogen is released (Figure 3.11). Stoichiometry and flame temperature are the most important factors governing the fractional conversion of fuel-N to NO_x, although in fuel-lean combustion, temperature dependency is significantly reduced. Under fuel-lean conditions, $[OH] \approx [O] \approx [H]$ and the following reaction scheme is assumed to take place:

$$HCN + O \rightleftharpoons NCO + H \qquad\qquad R\ 4.19$$

$$HCN + O \rightleftharpoons NH + CO \qquad\qquad R\ 4.20$$

$$HCN + O \rightleftharpoons CN + OH \qquad\qquad R\ 4.21$$

followed by
$$NCO + O \rightleftharpoons NO + CO \qquad\qquad R\ 4.22$$

$$NH + OH \rightleftharpoons NO + H_2 \qquad\qquad R\ 4.23$$

$$CN + O \rightleftharpoons N + CO \rightleftharpoons NO \qquad\qquad R\ 4.24$$

Under fuel-rich conditions, hydrogenated nitrogenous species are formed that may end up in the production of species such as NH_3 and N_2 according to

$$HCN + OH \rightarrow NH_2 + CO \qquad\qquad R\ 4.25$$

$$NH_2 + RH \rightarrow NH_3 + R \qquad\qquad R\ 4.26$$

$$NH_2 + NO \rightarrow N_2H \rightarrow N_2 \qquad\qquad R\ 4.27$$

There is also a greater probability of hydrocarbon radicals and carbon particles in the fuel-rich flame, depleting NO by

$$CH + NO \rightarrow HCN \rightleftharpoons HCNO$$

and
$$\qquad\qquad R\ 4.28$$

$$HCCO + NO \rightarrow HCNO \rightleftharpoons HCN$$

The correct prediction of fuel-NO formation within the flame requires coupling the NO formation kinetics to the hydrocarbon combustion mechanism in which the number of reactions to be considered can involve long computer processing times. As an alternative, a simplified mechanism of the form is currently utilized:

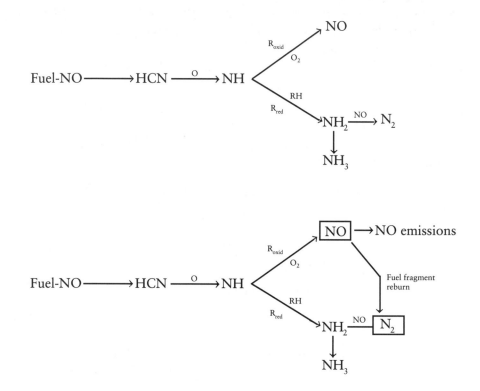

The postprocessing NO_x package used by the authors incorporates fuel-N being produced from both volatile and char-released nitrogen. In order to model NO_x formation accurately, the correct flame temperature, volatile release, char burn-out, and burner aerodynamics must be attained.

Fuel-Nitrogen Release during Coal Combustion

If the rate of release of nitrogen from coal volatiles (R_{VN}) is given by

$$R_{VN} = \frac{f_N \bar{S}_p}{M_N} \qquad\qquad \text{E 4.25}$$

and the overall mean reaction rate of hydrogen cyanide formation from volatile combustion is

$$\bar{R}_{HCN} = (R_{VN} - R_{NO} - R_{N_2})M_{HCN} \qquad \text{E 4.26}$$

then, the overall net rate of NO formation from volatile and char combustion is

$$\bar{R}_{NO} = (\bar{R}_{NO} - \bar{R}_{N_2} - \bar{R}_{char})M_{NO} + R_{NO_{thermal}} + R_{NO_{prompt}} \qquad \text{E 4.27}$$

with the instantaneous reaction rate terms for

$$R_{NO} = \rho(1\times10^{11})\,X_{HCN}X_{O_2}^{b}\,\exp\left(\frac{-33.74}{T}\right)\Big/M_m \qquad \text{E 4.28}$$

$$R_{N_2} = \rho(3\times10^{12})\,X_{HCN}X_{NO}^{b}\exp\left(\frac{-30.227}{T}\right)\Big/M_m \qquad \text{E 4.29}$$

$$R_{char} = n_p\sigma_p(4.17\times10^{7})\,A_sP_{NO}\,\exp\left(\frac{-17.48}{T}\right) \qquad \text{E 4.30}$$

The rate of formation of fuel-NO$_x$ and fuel-NO$_x$ concentrations within the coal burner can be predicted.

In an investigation by Weber et al. (1993), computation of NO emission based on the above approach was carried out. These investigators concluded that in order to model flue emission of NO$_x$ during coal combustion, information from prediction should be extrapolated gently from rigorously verified predictions. It seems to be the only way to predict NO$_x$ emissions accurately. Recent studies have shown that unlike natural gas burners, it is not possible to scale NO$_x$ emissions from one size burner to another. The following section explains the process for volatiles and char combustion.

Developments to NO$_x$ Models

Volatile-released NO   There is interest in the fate of ammonia and hydrogen cyanide in a flame (Glass and Wendt, 1982). Reactions under investigation include:

$$R_{HCH} = \frac{d\,[HCN]}{dt} = A\exp\left(\frac{-E}{T}\right)[HCN][H_2O]/[H_2]^{1/2} \qquad \text{E 4.31}$$

$$R_{NO} = \frac{d[NO]}{dt} = A \exp\left(\frac{-E}{T}\right)[NO][NH_3]/[H_2]^{1/2} \qquad \text{E 4.32}$$

$$\frac{d[NH_3]}{dt} = \frac{d[NO]}{dt} - \frac{d[HCN]}{dt} \qquad \text{E 4.33}$$

where A and E are the exponential factors and activation energy respectively and depend on the type of the coal and its composition.

NO from char reaction

Work that is currently under investigation regarding char-bound nitrogen-NO reactions concentrates upon both NO formation and destruction by the char and CO, respectively. It is assumed that the char consists of C, N, and inorganic compounds. The conversion of char nitrogen to NO is assumed to occur by

$$C(s) + \tfrac{1}{2}O_2 \rightarrow CO \qquad \text{R 4.29}$$

$$C(s) - N + \tfrac{1}{2}O_2 \rightarrow NO \qquad \text{R 4.30}$$

A conversion factor for char-bound nitrogen may be calculated from

$$\eta = \frac{[NO]}{\{[CO] + [CO_2]\}(N/C)_{char}} \qquad \text{E 4.34}$$

NO formed initially may subsequently be reduced. The NO conversion is increased with decreasing NO/CO ratio, which may be obtained from

$$\frac{NO}{CO} = \left(1 - 2 f_{tp}\right)\left(\frac{N}{C}\right)_{char} \qquad \text{E 4.35}$$

where f_{tp} is a function of temperature and NO partial pressure.

NO reduction may be represented by the carbon-surface catalyzed reaction

$$2NO + 2(CO) \rightarrow N_2 + 2CO_2 \qquad \text{R 4.31}$$

or by reduction occurring at a free carbon site resulting in chemisorbed oxygen on the char surface:

$$NO + C_{fas} \rightarrow C(O) + \tfrac{1}{2}N_2 \qquad \text{R 4.32}$$

CO reduction of NO is greatest at low temperatures and the chemisorbed oxygen reacts to produce CO_2:

$$CO + C(O) \rightarrow CO_2 + C_{fas} \qquad \text{R 4.33}$$

With increasing temperatures, however, the rate of desorption of the chemisorbed oxygen to form CO increases with the net effect of reducing the overall chemisorbed sites and hence the importance of reaction R4.23:

$$C(O) \rightarrow CO \qquad \text{R 4.34}$$

These reactions may be used to derive a Langmuir-Hinshelwood model for NO reduction in the presence of CO (Chan et al., 1983):

$$-r_{NO} = \frac{k_{33} P_{NO} (k_{33} P_{CO} + k_{34})}{k_{32} P_{NO} k_{33} P_{CO} + k_{34}} \qquad \text{E 4.36}$$

The corresponding values for the rate constants are

$$k_{32} = 2.1 \times 10^{-1} \exp\left(-13.1 \times 10^3 / T\right) \quad \text{kmol} / \text{s m}^2 \text{ atm} \qquad \text{E 4.37}$$

$$k_{33} = 7.4 \times 10^{-4} \exp\left(-9.56 \times 10^3 / T\right) \quad \text{kmol} / \text{s m}^2 \text{ atm} \qquad \text{E 4.38}$$

$$k_{34} = 1.5 \times 10^{-2} \exp\left(-20.1 \times 10^3 / T\right) \quad \text{kmol} / \text{s m}^2 \qquad \text{E 4.39}$$

The overall formation of NO from the char-bound nitrogen can thus be represented by

$$R_{NO} = \frac{-N_{char}}{C_{char}} R_{char} \qquad \text{E 4.40}$$

where the char reaction rate R_{char} is determined by

$$R_{char} = 254 \ \exp(-21.6 \times 10^{-3} / T) P_O^n \qquad \text{E 4.41}$$

Sulfur Emissions

The ability to model coal oxidation systems so that SO_x or H_2S end products can be predicted is of importance for both system design and maintenance. The sulfur content of coal is generally classified as being either pyretic or organic. During coal formation, pyretic sulfur is assumed to have been derived from sulfates and iron in the deposition water, while organic sulfates were derived from the reaction of sulfate compounds with decomposing organic material. When predicting the sulfur emissions from coal combustion systems, one must therefore represent the release of sulfur from both forms, requiring different reaction pathways to be considered. In general, however, sulfur is released during pyrolysis as H_2S and is oxidized almost immediately to SO_2. The paths are:

Pyretic Sulfur: The decomposition of pyretic sulfur occurs at temperatures around 750K due to its relatively unstable structure. The products are a function of the local gas atmosphere such that in an oxidizing environment the pyrite forms hermatite according to

$$2FeS_2 + 5\tfrac{1}{2}O_2 \rightarrow Fe_2O_3 + 4SO_2 \qquad\qquad \text{R 4.35}$$

In reducing environments, a more complex mechanism occurs, which may be summarized as

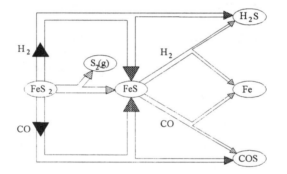

Reaction of FeS has a higher activation energy than that of FeS_2 and can therefore remain within the coal ash leaving the system.

Organic Sulfur. The release of organic sulfur during coal combustion has not been extensively investigated. Most work that has been performed previously concentrates on coal characterization with the disadvantages of assuming low heating rates and low maximum temperatures. Some results reported are of interest; i.e., Kelemen et al. (1991) show that during the initial stages of pyrolysis, aliphatic organic sulfur compounds are released prior to pyrites. The more stable aromatic organic sulfur compounds evolve after the decomposition of pyrites.

In order to model the formation of sulfur compounds within computational programs, it is therefore necessary to incorporate a number of different factors relating to:

1. The respective amounts of the different sulfur compounds within the coal, i.e., pyrites, aromatic and aliphatic organic compounds
2. The release of aliphatic organic compounds at lower temperatures
3. Reaction of pyrites and subsequent reaction of their products
4. Organic sulfur structure changes, i.e., from aliphatic to more stable aromatic structures
5. Reaction and possibly retention of aromatic structures within fly ash

The rate of release of H_2S can be estimated by computer models such as FG-DVC. The final products of sulfur compounds will be dependent upon the local gas environment, and thus coupling of oxygen concentrations with sulfur chemistry will be a necessity. In the simplest arrangement the H_2S can be considered to be converted to SO_2.

Interaction of SO_x on NO_x formation   The influence of fuel-sulfur on nitrogen oxide formation has been previously investigated. Although a number of different mechanisms have been proposed for sulfur interaction by various workers, the main consensus is that the presence of sulfur results in sulfur compounds such as SO, SH, H_2S, S_2 CS_2, and COS existing in superequilibrium in postflame regions. It has been suggested by Muller et al. (1979) that sulfur presence accelerates recombination of H radicals and those of O and OH atoms, thus resulting in reduced thermal-NO formation by the extended Zeldovich mechanism. The effect of fuel-NO formation is more complex and less well understood, although it is believed to be strongly dependent upon flame stoichiometry and flame temperature (Pfefferle and Churchill, 1989). Future developments to the NO_x prediction model will incorporate the influence of sulfur on overall NO_x formation by assuming that in fuel-rich zones, sulfur intermediates will react (Chagger et al., 1991) according to

$$SO_2 + H \rightarrow HS, H_2S \qquad \text{R 4.36}$$
$$HS + NO \rightarrow OH + NS \qquad \text{R 4.37}$$
$$2NS \rightarrow N_2 + S_2 \qquad \text{R 4.38}$$

Although the incorporation of these reactions into the NO_x package will result in NO_x reductions, the magnitude of these reductions is assumed to be small.

Modeling Reburn

Reburn is an in-furnace NO_x control technique previously described but one that now can be modeled by CFD methods. The amount of primary to secondary fuel dictates the staging ratio. The secondary zone forms CH and other related radicals that are responsible for the reduction of NO to N_2. A simplified reaction mechanism of the form

$$CH + NO \rightleftharpoons HCN + CO \qquad\qquad R4.39$$

can be used. The mechanism is complicated and involves many steps, some still disputed (Glarborg et al., 1998):

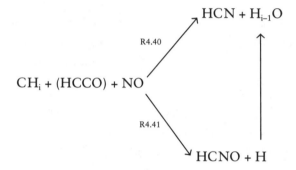

The HCCO species can be considered to be involved in a partial equilibrium:

$$HC + CO \rightleftharpoons HCCO \qquad\qquad R\ 4.42$$

HCN formed may then react to form N_2 via the reactions

$$HCN + OH \rightarrow HNCO + H \qquad\qquad R\ 4.43$$
$$HNCO + H \rightarrow NH_2 + CO \qquad\qquad R\ 4.44$$
$$NH_2 + NO \rightarrow N_2 + H_2O \qquad\qquad R\ 4.45$$

The efficiency of the reburn technique is determined by the initial stoichiometry of the primary zone. If the resulting oxygen concentrations are too high, then HCN formed in the reburn process may proceed to form NO. Furthermore, the injection of burn-out air after the initial reburn zone may result in NO_x formation via

$$HCN \xrightarrow{\ OH,O\ } NO + CO \qquad\qquad R\ 4.46$$

5

COMBUSTION MECHANISM OF COAL PARTICLES IN A FIXED, MOVING, OR FLUIDIZED BED

5.1
Fixed- and Moving-Bed Combustion

Coal can be burned in a bed where the grate is fixed in space and combustion takes place with an upward flow of air through the bed. The grates can be agitated or rotated to remove ash. Alternatively, the grates can be flexible and move, in which case this is called traveling-grate combustion. The use of the term is not clear-cut because it is sometimes used in fixed-bed situations where the bed itself is moving downward under the influence of gravity. Coal can be fed into these bed arrangements from above, called overfeed, or below, called underfeed, or indeed from any intermediate position.

Overfeed Burning   The fuel is fed as raw coal lumps onto the surface of the burning fuel bed. The new fuel is heated by the hot combustion gases, and care must be taken that the swelling nature of the coal does not cause blocking of the feed system. Figure 5.1(a) represents a bed of solid fuel burning on a fire grate through which a current of primary air passes upward to support the combustion. Industrial grates, in general, are assemblies of individual fire bars mounted parallel to one another so that a narrow gap separates neighboring bars through which the air current passes. The bars are arranged longitudinally with respect to the fire bed. If the fuel bed is burning brightly, a fresh change of coal is placed on it, the fresh coal is heated rapidly, and it decomposes with the evolution of volatile matter and the formation of a solid residue of coke (or char) and ash that falls through the grate. The upward current of air and the upward-moving, hot products of the combustion of the fuel bed cause the decomposition of the fresh coal charge to proceed from below upward. Ultimately, all the volatile matter is evolved and the fresh coal has been converted to coke and volatiles. The coke burns on the fire bed and the volatile matter burns in the space above the fire bed, known as the combustion space. The combustion of coal on a fire bed, supported on a grate thus involves these separated mechanisms:

1. The devolatilization of the fresh coal when placed on the hot fire bed
2. The combustion of the bed of the resultant coke and the formation of ash particles
3. The combustion of the evolved volatile matter above the fire bed. The processes are exactly the same as for pf combustion, except that because large particles are used the initial heating-up stage of the coal is longer and can involve extensive fragmentation of the initial lump of coal. The heating, mixing, and combustion processes are different from that of pulverized coal only in relation to their spatial arrangements; that is, the interparticle distances are small, and overall temperature is lower and generally uniform, although there may be hot spots.

The combustion of the bed of coke particles thus formed on the fire grate is maintained by the upward stream of air that passes through a layer of ashes on the bottom of the grate before it comes into contact with the coke. This air stream is known as the primary air supply.

As soon as the heated primary air meets the coke, the combustion reactions begin. The free oxygen in the air reacts with the coke and forms CO_2 and CO.

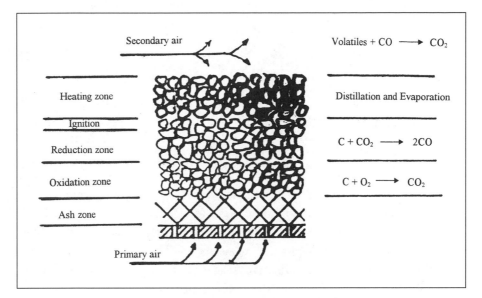

(a) Overfeed combustion

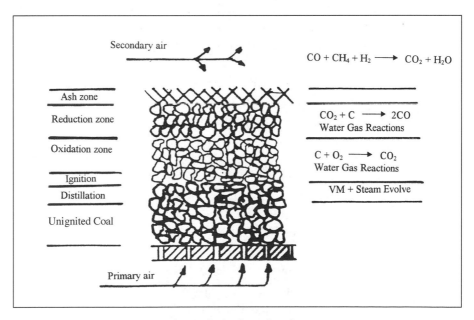

(b) Underfeed combustion

Figure 5.1   Modes of combustion on a grate.

The fundamental reactions are

$$C_s + O_2 \rightarrow CO_2 \qquad\qquad\qquad\qquad \text{R 5.1}$$
$$2C_s + O_2 \rightarrow 2CO \qquad\qquad\qquad\qquad \text{R 5.2}$$

where C_s refers to solid carbon. CO is oxidized with oxygen in the gas phase to give CO_2 that can react with the solid carbon:

$$C_s + CO_2 \rightarrow 2CO \qquad\qquad\qquad\qquad \text{R 5.3}$$

The rate of the carbon-oxygen reaction is far higher than that of the carbon–carbon dioxide reactions. At about 1100K the rate of reaction with oxygen is about 10^5 times that of reaction with CO_2, but at temperatures about 1300K the rate of CO production to CO_2 is about 10 to 20. At higher temperatures, such as in pf flames, the dominant reaction is to CO (R 5.2). Therefore an isolated carbon particle being at bed temperatures tends to produce mainly CO_2. This reaction is strongly exothermic, but all the oxygen in the air is consumed and thereafter reaction of the CO_2 occurs with the carbon as shown in Figure 5.1(a) to form CO.

The composition of the fuel bed can be divided into four reaction zones, with the zones overlapping one another. Figure 5.2 shows these zones and the results of investigations into the profiles of O_2, CO, and CO_2 through the bed. The formation of CO is undesirable, so stokers that work with a thin fuel bed, like that produced in the spreader stoker, have a definite advantage. The combustion products will leave the fuel bed at a temperature of about 1500°C.

It is necessary to control the process according to the demand for heat. This is achieved partly by adjusting the rate of supply of fresh fuel and partly by controlling the air supply. It is usual to find that the extent to which the fuel supply can be increased and complete combustion achieved is limited by the character of the fuel and by the design of the combustion apparatus. Equally, there is a limit to the extent to which the fuel supply can be reduced while continuous combustion is maintained.

The ratio of the maximum rate of satisfactory fuel consumption to the minimum rate of continuous consumption is known as the turndown ratio. This ratio is a function of the fuel and of the equipment.

Chain or traveling-grate burning (the stoker)   This combustion process takes place in a dynamic situation and is shown in Figure 5.3. In this arrangement the

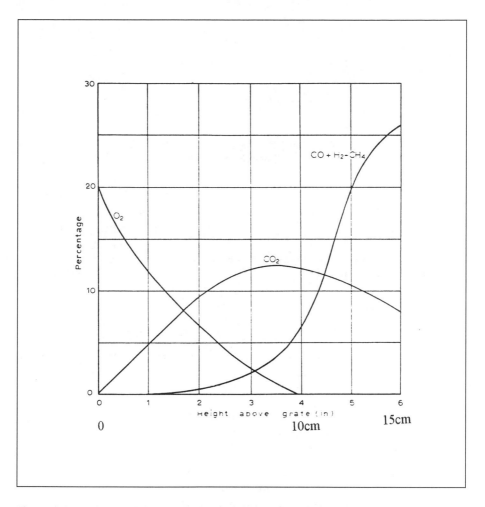

Figure 5.2 Variation of the composition of the combustion gases as they pass through a fire bed.

grate moves continuously from left to right, and as the fuel emerges from the hopper under the guillotine, the bed ignites from the top. The heat is supplied by radiation from the burning fuel and reflection from the front ignition arch.

To achieve the maximum rate of ignition, the air flow through the fuel bed at the front of the stoker must be limited, or the cold air would cool the fuel bed faster than it can be heated. The rate of feed and the stoker grate speed are affected by the rates of ignition and combustion. Various characteristics of the

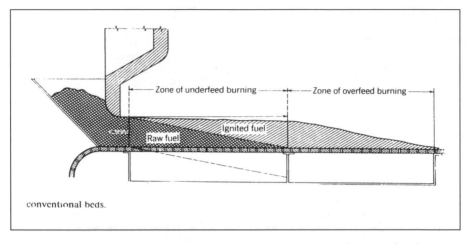

Zone of underfeed burning — Zone of overfeed burning —

Ignited fuel

Raw fuel

conventional beds.

Figure 5.3   Diagram of a traveling-grate stoker. For convenience, horizontal scale is shorter than in conventional beds.

coal, such as size, moisture content, volatile matter, swelling index, and coal rank, influence the rate of ignition. As the coal moves away from the ignition arch, the ignition front gradually moves downward, leaving burning fuel above. The volatile is emitted and burned, the coke burns to CO_2 as the fuel progresses, and finally, the ash bed is discharged over the rear of the stoker.

The spreader stoker also operates on the overfeed principle. Raw fuel is fed over the plane of ignition using a thin fuel bed. In the path of the flame and hot combustion gases the small coal particles burn in suspension. The larger pieces are ignited as they land on the burning coal bed. Ignition is extremely rapid. As the pieces of coal burn away, the ash is immediately chilled by the combustion air and remains granular. The traveling grate then dumps this ash into a hopper.

Underfeed burning   The term underfeed burning denotes that coal is fed from under the plane of ignition. As the raw coal reaches the plane of ignition it is heated and the moisture is driven off, followed by the evolution of volatiles. These gases then pass through the burning fuel bed, where the hydrocarbons mix with oxygen from the air. The compounds formed mix with more oxygen, and while passing through the burning coke bed are so well stirred that they burn with a relatively small amount of overfire air at full load — none at all at light loads.

Underfeed stokers operate with thick fuel beds. They are essentially coal gasifiers with gasified gases burning in the upper region and emit only a small amount of fly ash in the flue gas when properly operated. The processes that occur are shown in Figure 5.1(b). To illustrate the principle, Figure 5.4 shows the operation of a small underfeed stoker.

Conditions for complete combustion   It is essential to efficient and economical management of the combustion process that the secondary combustion over the fuel bed should be complete. To ensure complete secondary combustion it is necessary to

1. Have enough secondary air to burn all the combustible matter rizing from the fuel bed.

2. Mix the secondary air and the combustible matter intimately.

3. Keep the air/combustible matter mixture sufficiently hot to allow ignition to occur. If the temperature of the mixture falls below a critical value it will not ignite but will pass to waste unused.

4. Allow the secondary air and combustible matter to be in contact long

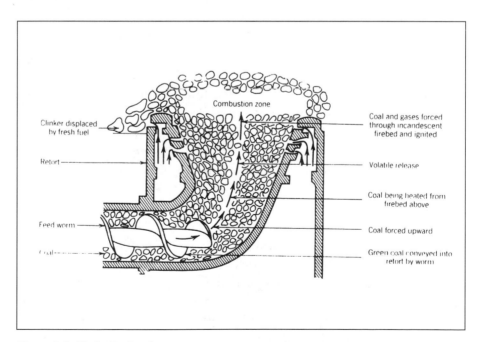

Figure 5.4   Underfeed stoker.

enough at a sufficiently high temperature to allow the secondary combustion reactions to be completed.

5. Prevent the burning mixture being chilled by contact with cold surfaces.

These conditions for complete secondary combustion apply not only to combustion above a bed of coal on a fire grate but to combustion systems generally.

5.2
Atmospheric Fluidized-Bed Combustion

The atmospheric fluidized-bed-combustion (AFBC) process consists of forming a bed of inert materials such as finely sized ash or ash mixed with sand, limestone (for sulfur removal), and solid fuel particles in a combustor and fluidizing it by forcing combustion air up through the bed mixture. It is thus significantly different from fixed-bed combustion because it is a better mixed and a better controllable combustor, but the air supply requirements and control equipment required makes it a more complex and expensive system. The atmospheric "bubbling-bed" type of AFBC technology (called bubbling fluidized-bed combustion, BFBC) has a defined height of bed material and operates at or near atmospheric pressure in the furnace. Generally, AFBC is used for heating, drying, and steam-raising applications, the last from in-bed boiler tubes.

The mode of operation is as follows. At low gas velocities, the gas flows through the bed without disturbing the particles significantly, and the bed remains fixed or slumped. As the gas velocity is increased, the force exerted on the particles by the upward passage of the gas increases until a point is reached where the gas stream supports the total weight of the bed. This marks the onset of fluidization and the gas velocity at this point is referred to as the minimum fluidizing velocity.

At gas velocities greater than the minimum fluidizing velocity, the excess gas passes through the bed as bubbles, and the bed is termed "fluidized." The passage of the bubbles through the bed gives the bed the appearance of a boiling liquid. This comparison is a good one because the resultant turbulent mixing of the particles provides the bed with some of the properties of a boiling liquid, including good heat and mass transfer and a hydrostatic head. The basic features of these beds and their development are described in a number of textbooks, e.g., Fluidized-beds (Howard, 1983), concerned with the development of the subject and the series of proceedings of the International Conference on Fluidized-bed Combustion.

The minimum fluidizing velocity and bed-particle terminal velocity are a function of the size of the bed particles, higher velocities being required both to fluidize and to entrain larger particles rather than smaller particles. The relationship between these velocities and the bed-particle size is illustrated in Figure 5.5 for conditions representative of a fluidized-bed combustor (particle density 2500 kg/m^3, air at 1 bar, and 850°C).

Conventional fluidized-bed combustors are operated at gas velocities that can be several times higher than the minimum fluidizing velocity and also several times lower than the bed-particle terminal velocity. Under these conditions most of the gas passes through the bed as bubbles that typically occupy 20–50% of the bed volume; the rest of the gas permeates through the bed material, i.e., the particulate phase. The passage of the bubbles through the bed causes agitation of the bed material. Vertical mixing is particularly effective because particles are carried rapidly upward through the bed in the wake of bubbles. There is a continuous interchange of gas between the bubbles during their passage through the bed and the particulate phase, resulting in good gas-solids reaction. When the bubbles reach the bed surface they burst, throwing showers of bed particles into the space above the bed. Some of these particles, especially the lighter ones, are carried out of the combustor entrained in the gas stream; this process is known as "elutriation."

As the gas velocity increases and the operating point moves from minimum fluidization toward total entrainment, the bubbling becomes violent and the bed expands because more of the bed volume is occupied by bubbles. The preferred operating conditions for a conventional fluidized-bed combustor lie in the relatively narrow range denoted as "good fluidization" in Figure 5.5.

In the mid-1970s the atmospheric "circulating" fluidized-bed-combustion technology (CFBC) was developed. CFBC has particular advantages, e.g., with respect to heat transfer, combustion efficiency, and fuel feed and these are enhanced by operating at elevated pressures.

A fast circulating fluidized-bed differs from the conventional bubbling bed in a number of respects:

1. The bed fills the containment vessel, and so the geometry is different; such beds are tall.

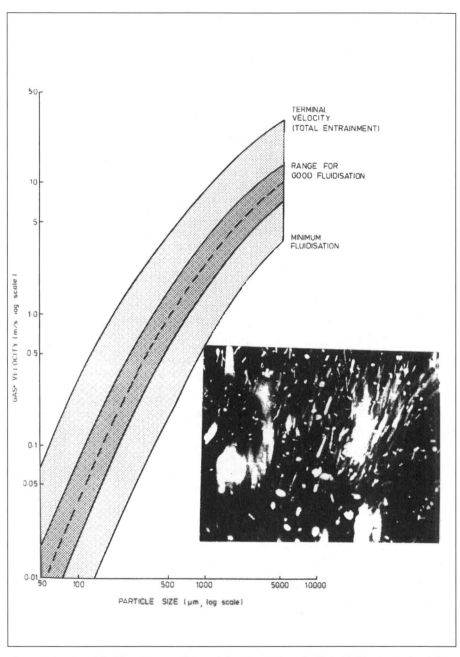

Figure 5.5   Good fluidization conditions (after Battock and Pillai, 1977). Insert is of the top of a correctly fluidized burning bed of coal particles.

2. There are no bubbles; instead, clusters and strands of particles are turbulently mixed with the gas stream and are continuously breaking up and reforming.
3. The density of the bed depends on the flow rate of recycled solids. If it is reduced, the system changes in character from a fast fluidized bed to a dilute phase system.

Modeling an Atmospheric Fluidized Bed

The modeling of combustion in a fluidized bed presents a considerable number of problems because of the complex hydrodynamics. Modeling really consists of two parts: (1) modeling the bed itself and (2) modeling the coal combustion processes. The fluidized-bed model is based on the concepts described above, and models of coal combustion/gasification in fluidized beds commonly utilize the two-phase fluidized-bed model. In this particular model, gas in excess of the minimum fluidizing velocity is assumed to flow through the bed as bubbles, while the emulsion remains stagnant at fluidizing conditions. Experimental findings, however, illustrate that this type of model produces acceptable results only under certain conditions, due to the simplistic approach used to represent both the emulsion and bubble movement within the bed. To overcome these shortcomings, it is assumed that the wake behind each bubble can result in the transport of solids up through the bed with a corresponding recirculation of solids down through the emulsion phase. This approach, which is commonly known as the K-L Model with Davidson Bubbles and Wakes, provides a more accurate representation of the bed (Kunii and Levenspiel, 1991).

Bubbles entering the bed expand as they pass up through the bed. Each bubble is assumed to consist of a bubble volume that is surrounded by a bubble "cloud." The size of this cloud depends on the bubble velocity and the minimum fluidization velocity. Following the bubble is the bubble wake, which may entrain solids up through the bed. Recirculation of these solids can occur within the emulsion phase, and in systems with large internal bed recirculation, such as the ABGC, the gas velocity through the emulsion can become negative and flow down through the bed. Transport processes occur between the bubble phase, the cloud, and the emulsion. General fluidizing equations are solved by splitting the reactor or fluidized bed into a number of compartments and solving the equations representing coal conversion, mass and heat transfer, etc., within each compartment.

The coal combustion model is based on the same principles outlined in the pf model. Under normal operating conditions the volatile matter is released as the coal particles heat up and will burn completely. Some of the combustion will occur above the bed because part of the volatile matter is inevitably released in a zone where oxygen is not available in the correct amounts, near the coal feed point and at the top of the bed.

The reaction in the fluidized bed between oxygen and the char particles remaining after the removal of the volatile matter is constrained by three mechanisms:

1. Transfer of oxygen from the bubbles to the particulate phase
2. Diffusion of oxygen to the particle surface and of carbon dioxide from the particle surface
3. Chemical reaction at the particle surface

For large char particles the rate-controlling step is usually the diffusion of carbon dioxide from the particle surface (mechanism 2). With fine particles, however, diffusion rates are high and chemical reaction rates (mechanism 3) become controlling. Typically, the changeover from diffusion to chemical rate control occurs at particle sizes in the range 50–100 μm. Since chemical reaction rates increase more with temperature than diffusion rates do, the changeover occurs at a smaller particle size with a higher bed temperature and vice versa. The rates of reaction of the oxygen with the char particle (and the volatiles) are exactly the same as for pulverized particles described in the preceding chapter except that the temperature is lower and in some cases the pressure is higher. In addition, are the effects of the bed on char fragmentation and erosion of ash from the char-particle surface.

The bed mainly consists of noncombustible inert material such as sand, the carbon concentration in the bed being generally less than 1% if crushed coal is used and less than 5% for larger size gradings. These values provide sufficient surface area for the rate of reaction to be consistent with the rate of oxygen supply. It is the low carbon concentrations in the bed that makes a fluidized-bed combustor tolerant to the caking properties of the coal. The average temperature at the surface of the burning char particles is up to about 100°C higher than the bed temperature, and this permits the necessary high rate of heat transfer from the char to the bed.

As the char particles burn they reduce in size until some become entrained in the gas stream and are removed from the combustor by elutriation. The size of the entrained char particles is mainly between 10 and 100 μm, so that most of this material is able to be collected using cyclones. Some loss of char can also occur with the removal of excess bed material to control the bed depth, but because of the low char concentration in the bed this loss is small compared with that by elutriation. The total char loss typically represents 2–15% of the thermal input of coal to the combustor. Values above 4% are unacceptable commercially, and the design must incorporate features to improve the combustion efficiency to an acceptable level.

In general, not all of the char combustion occurs in the fluidized bed. As bubbles burst at the bed surface, bed particles are thrown into the space above the bed, the freeboard, and combustion of the char continues in this region. A proportion of the heat released by combustion of both char and volatiles in the freeboard returns to the bed as sensible heat in the particles that fall back. The remainder is responsible for heating the combustion gases above bed temperature by as much as 200°C in some systems. In pressurized combustors, where the gas is cleaned before cooling, some combustion may occur in the cyclones.

Sulfur retention Fluidized-bed combustion can control gaseous emissions of SO_2 and NO_x during combustion by the addition of limestone or dolomite (SO_2) and through low combustion temperatures and staged combustion (NO_x). AFBC is a very suitable combustion technology for biomass, waste materials, and low-grade fuels (such as coal recovered from coal cleaning operations) or for mixtures of them, termed cofiring.

The added limestone or dolomite retains the sulfur released as calcium sulfate that can be removed with the ash. The course of the reactions involved depends on the operating conditions and in particular on the pressure.

At atmospheric pressure, limestone and dolomite calcine (decompose) according to the reactions

$$CaCO_3 = CaO + CO_2 \qquad\qquad \text{R5.4}$$
$$CaCO_3 \cdot MgCO_3 = CaO \cdot MgO + 2CO_2 \qquad\qquad \text{R5.5}$$

The calcium oxide, which has a porous structure, then reacts with sulfur dioxide to form the sulfate, as follows:

$$CaO + SO_2 + \frac{1}{2}O_2 = CaSO_4 \qquad \text{R5.6}$$

$$CaO \cdot MgO + SO_2 + \frac{1}{2}O_2 = CaSO_4 \cdot MgC \qquad \text{R5.7}$$

The kinetics of these reactions are well studied and depend on the porous nature of the CaO — the way in which the pores in freshly prepared CaO close up with time (aging) and the blocking of the pores by the formation of the reaction products. At high pressures, calcium carbonate is stable at the temperatures used in fluidized-bed combustion and does not calcine. As a result, the calcination of dolomite takes the form

$$CaCO_3 \cdot MgCO_3 = CaCO_3 \cdot MgO + CO_2 \qquad \text{R5.8}$$

Limestone and partially calcined dolomite react with sulfur dioxide according to the equations

$$CaCO_3 + SO_2 + \frac{1}{2}O_2 = CaSO_4 + CO_2 \qquad \text{R5.9}$$

$$CaCO_3 \cdot MgO + SO_2 + \frac{1}{2}O_2 = CaSO_4 \cdot MgO + CO_2 \qquad \text{R5.10}$$

Magnesium sulfate does not form when dolomite is used as the sulfur acceptor, either at atmospheric pressure or at high pressures. This is because magnesium sulfate is not stable at fluidized-bed-combustion temperatures.

For operation at high pressures, limestone is found to be unreactive and less effective. Since limestone does not calcine at high pressures, the porosity of the stone remains low and the gas cannot penetrate easily to the interior of the particles. After the formation of a surface layer of sulfate the reaction effectively ceases because no more active sites are accessible. When dolomite is used, the calcination of the magnesium component opens up the pore structure so that the reactivity of the stone is high and sulfation is not limited to the exterior surface of the particles.

As a consequence, limestone is the preferred sulfur acceptor at atmospheric pressure, and dolomite at high pressures. At intermediate pressures of up to about 4 bars, but depending also on the excess air level, some calcination of calcium carbonate will take place and both types of acceptor are equally effective.

Acceptor regeneration   In most fluidized-bed-combustor designs that incorporate sulfur retention, the limestone or dolomite is used on a "once through" basis so that reacted limestone is disposed of after removal from the combustion system. This can cause problems with the utilization of ash for construction purposes or indeed its disposal in landfill. In order to reduce the environmental impact of disposing of acceptor materials and also to save the associated costs, various schemes for regenerating the reacted limestone to the oxide or carbonate have been proposed. The main options are as follows.

1. *Thermal decomposition.* At temperatures above about 1000°C, calcium sulfate decomposes to give calcium oxide. This reaction could, for example, take place in a bed fluidized by air and in which sufficient fuel is burned to maintain the required temperature. Subsequent processing would then be required to remove the sulfur dioxide from the combustion gases, although the gas volume involved is comparatively small.

2. *Reduction.* Calcium sulfate can be reconverted to calcium carbonate (which may calcine to calcium oxide) using carbon monoxide. A number of reactions are possible, for example,

$$CaSO_4 + CO = CaCO_3 + SO_2 \hspace{2cm} \text{R5.11}$$

where the carbon monoxide required has to be generated externally (e.g., by a gasification process) and the operating conditions have to be chosen carefully.

Bed composition   Usually, the mineral matter present in the coal can be regarded as an inert material in the combustion process. However, some chemical changes do take place and usually result in a weight loss, for example, carbonates and sulfides decompose to form oxides. The residue after these changes is referred to as ash, as in the case of pf combustion.

The net effect of the combustion and sulfur, retention reactions is to provide a bed consisting of the following components: (1) coal present mainly as particles of low volatile char, accounting for less than 1% of the bed material if crushed coal is used and less than 5% for lump coal; (2) limestone or dolomite present mainly as calcined and partially sulfated stone; (3) coal ash; and (4) inert additive, for example, sand.

Commercial Development of Fluidized-Bed Combustion

Atmospheric fluidized-beds have passed the research and development stage, and AFBC units have commercially been available for about 25 years; there are many hundreds of units installed worldwide, especially in China. AFBC concepts with capacities of up to 200–400 MWt are now considered to be a commercial technology for power generation and industrial applications.

In order to make the AFBC technology even more attractive, a number of issues require further development, namely: fuel feed and ash handling for off-design feedstock to achieve the proposed fuel flexibility, predicting performance with respect to agglomeration and deposition, optimization of emission control, operating parameters and sorbent feed, utilization of solid residues (bed material and fly ash), and application to co-combustion of biomass and waste on a commercial scale There is a further issue that is of significance. Because of the bed temperature, some of the NO produced is reduced by the carbon to N_2O. Consequently, while NO emissions are low, the N_2O emissions are relatively high, often 100–200 ppm. These can, however, be reduced by use of an afterburner in the flue gas.

AFBC technology is expected to continue to play an important role in the intermediate market, especially for low-rank fuels—with capacity demands of possibly up to 500 MWt. However, large-scale applications of the AFBC technology have to be demonstrated and at the present time they have to compete with the much cheaper pf technology, especially from supercritical electrical power generation. However, recent economic analysis show an interesting balance between the technologies.

In the field of AFBC, further research and demonstration are necessary to make this technology less expensive and more reliable. In addition to short-term research, most of these research topics may well be investigated and demonstrated using existing AFBC units. Improvements achieved may be incorporated directly into the next generation of commercial-scale plants and thus be demonstrated within the next 5–10 years.

5.3
Pressurized Fluidized-Bed Combustion (PFBC)

A PFBC system operates a fluidized bed at an elevated pressure. Because of the higher pressure, the exhaust gases from PFBC can have sufficient energy to power a gas turbine, while the steam generated in the in-bed boiler tubes drives a steam turbine. This combined cycle configuration allows net efficiencies in excess of 45%.

In a similar way to AFBC, PFBC can control gaseous emissions during combustion by addition of limestone or dolomite (SO_2) and through low combustion temperatures and staged combustion (NO_x). For gas turbine applications, clean flue gases are necessary and there is a need for high-temperature particulate removal systems. Cyclones have been used for a coarse particulate removal upstream, and an electrostatic precipitator or filter is required downstream of the economizer to remove the remainder of the fly ash.

Generally the gases have to meet the following specifications (Mitchell, 1998) for a gas turbine:

TABLE 5.1
GAS TURBINE SPECIFICATION*

Contaminant	Chemical formula	Emission limit [†]	Comment
Particulates		2 ppmw	Grain size: >10 μm: 0; 2–10 μm: 7.5%; 0–2 μm: 92.5%
Sulfur compounds	$H_2S + COS + CS_2$	20 ppmv	Emission only.[‡] Corrosive only in combination with alkalis
Nitrogen compounds	$NH_3 + HCN$	—	Emission only
Hydrogen halides	$HCl + HF$	1.0 ppmw	
Alkalis	NaK	0.03 PPMW	
Heavy metals	V	0.05 ppmw	
	Pb	1 ppmw	
Calcium	Ca	1 ppmw	

*Specifications for Siemens Model VX4-3A gas turbine.
[†]Based on a lower heating value of 4 MJ/kg fuel, typical of fuel gas diluted with nitrogen and steam for NO_x suppression.
[‡]In order to reach very low SO_2 content in the flue gas of about 25 mg/m³ [6% O^2, STP (0°C, 273K), 101.3 kPa].

The development of PFBC has been underway since 1969, but still today, the PFBC technology is in the early stages of commercialization. Five PFBC units of less than 80 MWe, two in Sweden (Värtan), one in Spain (Escatrón), one in the US (Tidd), and one in Japan (Wakamatsu), have been put into operation.

This can be regarded as the "first generation" PFBC technology. The "second generation" PFBC technology may utilize a topping combustor to increase the inlet temperature to the gas turbine. In this case, a device for high-temperature, high-pressure particulate removal (HGCU) has to be installed between the fluidized-bed combustor and topping combustor to remove virtually all the ash upstream of the topping combustor (Mitchell, 1999). Due to the high gas turbine inlet temperature, significant additional cycle efficiency can by achieved, resulting in a net thermal efficiency of some 50%.

In terms of operational behavior and primary emission control, circulating PFBC technology may have advantages over bubbling-PFBC technology. The following issues still require further research: gas turbine operation in a "high-dust" environment, improvements in the overall reliability, availability and maintainability, and improved and simplified plant design resulting in reduced capital expenditure, and thus, reduced cost of electricity.

Corrosion due to volatile alkali species restricting the use of feedstock with high alkali metal or high chlorine content is a major problem still, and there is a need for sorbent technologies. Methods for reducing N_2O emissions are necessary just as much at high pressures.

For the future development of PFBC technology, a number of issues require further development, namely: topping combustor technology, efficient high-temperature (800–900°C) particulates removal systems, and advanced rugged gas turbine technology for high-temperature flue gas.

PFBC technology is under demonstration today. Even though this stage is well under way, further research and demonstration work are necessary to improve particular components and operational performance of PFBC systems (e.g., gas turbine operation in a high-dust environment, improved and simplified plant design resulting in reduced capital expenditure, proper feedstock preparation, and emission related issues).

As soon as sufficient progress in the combustor technology (e.g., circulating PFBC), hot gas cleanup (HGCU), and advanced gas turbines has been made, demonstration projects are appropriate to verify the significant potentials of the next generation PFBC technology.

5.4
Pressurized Circulating Fluidized-Bed Combustion (PCFB)

The pressurized circulating fluidized bed was devised to overcome a problem of fluidized combustion in which, as the ash is removed, some unburned carbon is removed with it and there is a loss of efficiency.

Combustion efficiency in PCFB is about 90–95% and overall plant efficiency when used in a combined cycle system is high. From a combustion point of view a number of interesting questions are posed and considerable research has been undertaken, e.g., MacNeil and Basu (1998). The difference between a fast circulating fluidized bed and a bubbling or a fixed bed is that the former bed is highly expanded and there are different interactions between the burning coal particles, and of course there is the effect of pressure on devolatilization of coal and char burn-out that is still not well understood.

The overall process of combustion, including devolatilization and char burn-out, therefore follow the rules set out in Chapter 4 with corrections for pressure. Effectively at these temperatures char particles burn under kinetically controlled conditions. The kinetic parameters vary as the pressure is changed; the surface reaction rates increase with pressure up to a pressure of 5 bars, and further increase leads to a decrease.

A number of research plants are operating burning coal, sewage, sludge, or biomass. There are a number of commercial plants in Europe, the US, and Japan. These are discussed in Chapter 6.

6

INDUSTRIAL APPLICATIONS OF COAL COMBUSTION

Coal combustion is used for a wide range of applications, especially industrial plant such as boilers or furnaces and for heating certain materials directly. An outline of these applications is shown in Figure 6.1. The major application of pulverized coal is for electricity production, as previously described, but another major application is for cement manufacture, where the pulverized coal is used to directly heat the cement feedstock (limestone) and the residual ash is incorporated in the cement powder, and of course, industrial heating forms another important sector.

In general, pulverized-fuel, fluidized-bed, or fixed-bed combustion are used to generate steam by indirect heating where the ash cannot contaminate the product. Coal combustion can be used where the gaseous combustion products, i.e., the hot gas, are used to power a gas turbine, but this type of gasification (complete combustion) is used together with hot gas cleanup to give clean gaseous products. This application is described in the next chapter. Pulverized coal has been used to power gas turbines and diesel engines (as proposed by Diesel) directly, but also without success — or at least so far.

In all these applications coal has a number of attractive features, the major ones being the relative ease of handling and storage of the fuel, and the highly

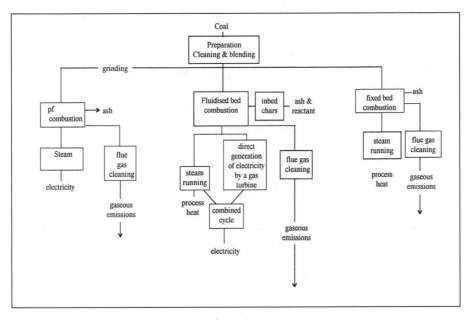

Figure 6.1 Industrial applications of coal combustion.

radiative nature of the flames produced that can be advantageous in certain applications.

6.1
Coal Storage, Preparation, and Blending

Coal size is largely determined by the method of mining and/or preparation. It is usually stored and transported in small pieces 5–10 cm in diameter and the storage is in silos, hoppers, or on a large scale, in stock piles and is normally stored prior to use. Storage procedures have been developed to prevent spontaneous combustion during storage, but nevertheless, slow oxidation takes place during storage with consequential loss of fuel. Care is taken to prevent ingress of air by compacting stock piles, etc., or avoiding the use of coal liable to spontaneously ignite. If pulverized coal is used for combustion it is pulverized immediately prior to use; pulverized fuel can be stored and transported, but extra precautions have to be taken to prevent spontaneous ignition and fire or explosion. Domestic and small-scale commercial or industrial applications of coal have for many years been based on the use of fixed- or moving-bed combustion or small-scale fluidized-bed combustion. The nature of the application determines the level of coal preparation that is undertaken, as does the source of the coal, that from a deep or opencast mine, or the quality of the coal seam itself.

Prior to cleaning, coal may be sized by the use of screens. The cleaning process currently involves, depending on the circumstances, some aspect of mechanical cleaning, and washing or wet cleaning. Much of this process has been directed in the past to removing the inorganic mineral matter, which is often referred to as dirt. This removal process has the advantage of removing some of the sulfur-containing mineral, namely, pyrite, from the coal. In some countries the pyrite content is massive and can be readily removed by normal coal preparation methods, but in others, e.g., the UK, much of the pyrite is finely disseminated and difficult to remove. Pyrite can be removed by a number of methods set out in Table 6.1.

Coal Cleaning and Blending

Coal cleaning and upgrading are used to reduce ash content, and this process is used in conjunction with the blending of coals to meet desired commercial specification. Generally, the blending of coal results in the additive effects of some of the properties of the coals being blended; for example, the sulfur and the fuel-N content is additive. Some features, such as carbon burn-out and slagging are not additive, however, and follow complex laws. Coals are often blended to meet commercial specifications, and typically these involve sulfur content, ash, volatiles, and calorific value.

TABLE 6.1
PHYSICAL PROPERTIES OF COAL AND THE MINERAL MATTER IMPURITIES

Physical property	Organic coal	Mineral (exc. pyrites)	Pyrites	Separation method
Density, g/cm^3	1.15–1.5	2.4–3.9	4.8–5.0	Density separation
Contact angle, degrees	49–68	13	58–73	Froth flotation
Magnetic susceptibility, 10^{-6} emu/g	−0.42–0.77	15–45	5–120	Very strong magnetic fields
Resistivity, ohm m	106–1011	102–106	1–102	Electrostatic separation

6.2
Combustion on Fixed or Traveling Beds

The Industrial Revolution in the mid-nineteenth century was based on coal combustion, and the use of fixed-grate combustion was central to the technology. The major part of the coal used was for raising steam in boilers or for heating metal (and of course, for making iron and steel in the blast furnace, which is not described here). The advent of oil and gas confined coal applications mainly to steam-raising using fixed or moving grates of one type or another. Steam boilers (or more correctly, steam generators) consist of two main classes known as shell-type boilers or water-tube boilers, briefly outlined in the following section. They have been widely used for fixed-grate firing in the past and indeed in some countries now; they can be used with other forms of firing.

Types of Shell Boiler

There are two types of shell boiler, the Lancashire and the Economic, that cover the range 3–20 MW. These are described next (Macrae, 1996).

Lancashire   The Lancashire boiler consists of a large horizontal cylindrical vessel (the "shell") that contains both the water and the steam, with provision for pipes to remove the steam as needed and to supply fresh water to be evaporated [Figure 6.2(a)]. The shell is fabricated from mild-steel plates. Two large-diameter, twin, steel tubes traverse the shell longitudinally. Each of these furnace tubes contains at its front end a fire grate, and the whole tube can act as a combustion chamber; they can be fired by a grate or pf and the internal furnace diameter is a function of the technique used (i.e., the combustion intensity). The hot combustion products transfer some of their heat energy to the water through the furnace-tube wall. More heat is transferred from these gases to the water through the lower part of the shell by causing the hot gases to travel in a longitudinal flue with brick walls from the back to the front below the shell and to return to the back of the plant via side flues, again of brick. The hot gases then pass to the chimney and some of the heat generated by combustion is lost; the boiler efficiency is about 78%.

Economic   The Economic boiler, shown in Figure 6.2(b), achieves a higher level of heat transfer than the simple Lancashire boiler. In it, because of its re-

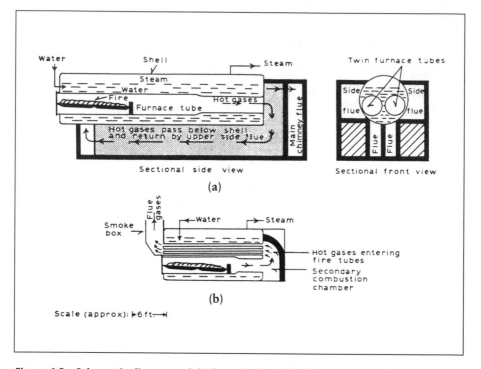

Figure 6.2   Schematic diagrams of shell-type boilers. (a) The Lancashire boiler, (b) the economic boiler of the same rating as (a)

duced length, the secondary combustion may not be completed before reacting gases leave the main furnace tube. A further combustion chamber, often brick lined to avoid undue chilling of the gases, is provided at the rear of the setting. The hot gases then pass through longitudinal steel tubes (the fire tubes) immersed in the water in the shell; these provide a high heat transfer surface to the water. They also cool the combustion gases to such an extent that any secondary combustion reactions not completed in the combustion chambers are effectively stopped. This may cause the deposition of carbon in the system, and indeed, the collecting space into which the waste gases pass from these fire tubes prior to the stack is named the smoke box

All modern shell-type boilers have an ancillary unit, the economizer, intended to remove still more sensible heat from the hot waste gases before they finally pass to the chimney. An economizer is an arrangement of steel or cast iron tubes through which the incoming water (the feed water) flowing to the boiler must pass and around which the still-hot flue gases flow on their way to the chimney. The feed water is consequently preheated and the waste gases are cooled. The economizer can suffer acid corrosion if the temperature of the outer surfaces of the tubes is allowed to fall below the dew point of the flue gas. A limitation is thus imposed upon the effectiveness of an economizer. This limitation is a function of both the sulfur content of the fuel and the combined effect of its water and hydrogen contents in determining the moisture content of the flue gases.

Water-tube boilers In the second main class of steam generator the water is circulated in a number of steel tubes around which the hot combustion gases flow. This is the class of water-tube boilers. The water tubes are mainly arranged in parallel, vertical planes and are themselves vertical or somewhat inclined. They are arranged in banks of tubes whose disposition around the burning fuel and in the path of the hot gases flowing toward the chimney flue allows them to receive heat both by direct radiation from the combustion zone and by transfer of the sensible heat of the combustion products. High efficiencies of heat recovery and high levels of superheat in the steam are possible by this arrangement.

Such boilers can easily be combined with a traveling-grate-type of grate as shown in Figure 6.3. This makes them suitable for boilers in the range of 0.5–20 MW, and the boiler design with slight modification makes them suitable for pf firing for power stations as described later in this chapter. It is possible to undertake mathematical modeling of this type of furnace as described later.

6.3
Conventional and Advanced Pulverized Coal-fired Boilers

Pulverized coal-fired boilers have been in use since the early 1900s and are currently the most widely accepted technology for large-scale coal-fired heat and electricity generation. A typical power station is shown in Figure 6.4. They are termed coal-fired boilers or just pf boilers.

Most of the conventional PCF boiler systems currently in operation use subcritical pressure < 221.2 bars) steam cycles with superheated and single reheated steam. This results, depending on feedstock, steam conditions, condensing pressure and plant size, in thermal efficiencies in the range of 35–38 % (based on the LHV).

A smaller number of units already operate with supercritical steam cycles

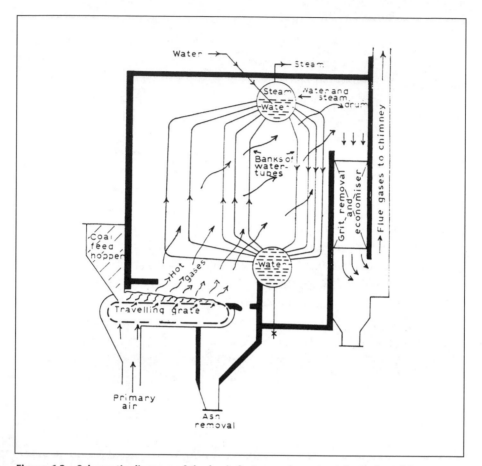

Figure 6.3   **Schematic diagram of the basic features of a water-tube boiler with convective circulation.**

(steam pressure some point above 221.2 bars, single reheat and main steam and reheat steam temperature around 540°C) that, together with some other means of thermodynamic optimization and an increase of plant capacity, raise the efficiency to up to 44%. In certain countries having coastal sites with access to cold sea water, such as Denmark, the efficiency can be raised to 48% for pf plants. Higher efficiencies can be obtained by further raising steam parameters to the "ultra-supercritical" conditions with maximum steam pressure above 248 bars and maximum steam temperature above 566°C. The efficiency of certain plants using high-moisture low-rank solid fuels (e.g., brown coal) can also be increased by applying external drying processes.

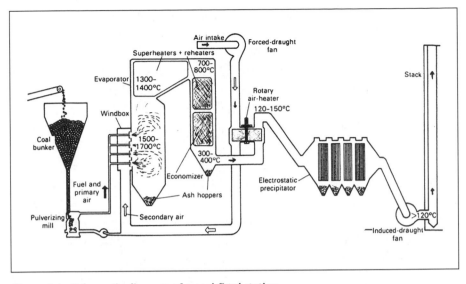

Figure 6.4   Schematic diagram of a coal-fired station.

In addition to the thermodynamic improvements, low-price emission control technologies (e.g., for SO_2, NO_x) have to be developed and commercialized in order to strengthen the economic competitiveness of advanced pf combustion. Thus, in addition to these, any advance in pf combustion systems requires the following: (1) advanced understanding of the combustion process including carbon burn-out and the formation of pollutants, and low-price primary or secondary emission control (e.g., SO_2, NO_x); (2) advanced computational methods to design combustion chambers and associated understanding of corrosion, erosion, and slagging; (3) commercial large-scale demonstration of supercritical steam cycles together with the appropriate steam turbine technology.

6.4
Equipment for the Combustion of PF in Power Plants: Burners and Furnaces

General features of industrial burners   The essential components of an industrial burner are: (1) a burner to produce the flame, (2) an air register that admits the air and determines the flow pattern to promote good mixing and NO_x control, (3) a burner throat designed to ensure stable combustion, (4) a control system for ignition and flame detection and determining the firing rate, together with a method of controlling the fuel/air ratio; all these must be a correct match with the combustion chamber.

A simple burner consists of a central tube through which pulverized coal and air are passed at the appropriate stoichiometry surrounded by a supply of secondary air that is normally swirled to stabilize the flame. Oil burners are often used for the a light-up of coal-fired boiler plant and may be used as part-load carrying burners in a dual coal-oil-fired plant.

Low-NO$_x$ burners are exclusively being used for all modern fossil fuel-fired power stations. In a coal-fired or dual coal-oil-fired power station, two types of burner arrangements are used: (1) linear staged flames where the rich zone is followed by a lean zone and (2) corner (tangential) firing. In the first type, axial swirl may be used to provide the primary reaction zone with secondary or tertiary air provided on the outside; a typical burner is shown in Figure 6.5. This type of low-NO$_x$ burner is commonly used for coal but can be used for coal/oil dual-fuel systems; the oil gun used for light-up and for oil firing would be centrally positioned. In atomizers of this type the jet can be shut off when not in use, and would have tip recirculation so that it is self-cooling and the fuels are available at the correct firing viscosity and pressure. Such burners are widely used, but there are many types of burners in use.

Pf burners can range from single burners as small as 50 kW or as large as 50 MW and can be mounted in arrays in a combustion chamber with an overall firing capability of about 600 MW.

Steam boiler applications   The major application here is for large water-tube boilers in thermal power stations, but there are countless applications in smaller steam-raising plant and central heating complexes.

The boilers are specially designed to match the heat transfer characteristics of the flames used and employ a first section designed for radiant heat transfer second stage.

In thermal power stations the combustion chamber is essentially rectangular in shape, and a number of burners may fire from one wall, or from opposing walls or even from the corners; this last case is termed tangential or T-firing. Typical combustion chamber configurations are shown in Figure 6.6, but the burners have to be chosen carefully to match a particular combustion chamber. The split flame burner that is illustrated in Figure 6.5 incorporates linear and also an element of parallel jet staging. These burners are normally used in wall

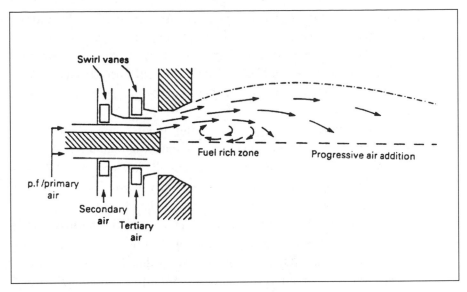

Figure 6.5   Typical arrangement of a low-NO$_x$ burner.

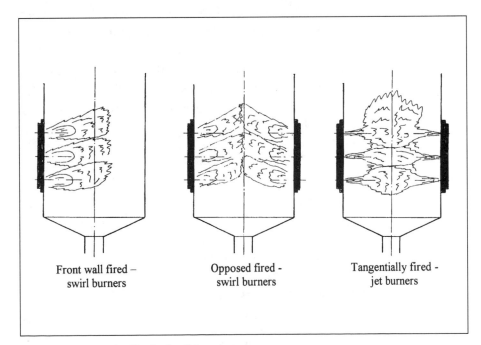

Figure 6.6   Typical utility boiler firing patterns.

firing or opposed firing using pf as the fuel, but can be used with a substantial amount of oil firing. While Figure 6.5 seems to indicate that the burner construction is simple, this is not the case. Figure 6.7 shows a typical arrangement in a practical burner, and the complexity in providing stratified and split flames is apparent. Figure 6.8 shows typical burners wall firing in a power station. Figure 6.9 shows tangential firing.

The third type is used for corner- (tangentially) fired boilers for coal (oil or dual-fuel) operation. The flame configuration is shown in Figure 6.6(c) and in Figure 6.9, and the staging between primary air and the fuel stream and the secondary air stream is apparent. The gas in the fuel-rich central zone has to pass through a lean region before leaving the combustion chamber.

In addition to the burners shown, complete NO_x control systems may invoke flue gas recirculation if appropriate, air staging, or reburn.

The operation to reduce NO_x must not increase the amount of unburned carbon. Consequently, the combustion technology employed must be taken into account for the characterization of the coal. The strategy often used is to (1) maximize the site of volatile evolution and its total yield from the fuel: this implies a high heating rate to as high a temperature as possible; (2) provide an initial rich zone to minimize NO_x formation, but there must be sufficient oxygen present to stabilize the flame and to enable reaction of HCN, NH_3 to proceed to N_2; (3) optimize the reaction time and temperature under fuel-rich conditions, i.e., maximize N_2 formation; (4) maximize the time the particles devolatilized char/char spends under rich conditions to remove as much fuel-N from the char as possible and convert that to N_2; (5) add sufficient staged secondary and tertiary air as possible to complete the burn-out.

6.5
General Features of Combustion and Furnace Design

Combustion stoichiometries and flame temperatures   If the elemental composition of a hydrocarbon is known, it is possible to calculate the overall stoichiometry of the combustion process. In general, for combustion of a hydrocarbon C_uH_v with air we have the overall reaction

$$C_uH_v + zO_2 + 3.76\,zN_2 = uCO_2 + \frac{v}{2}H_2O + eO_2 + 3.76\,zN_2$$

where $z = u + v/4 + e$, e is the number of moles of excess air, and 3.76 is the ratio

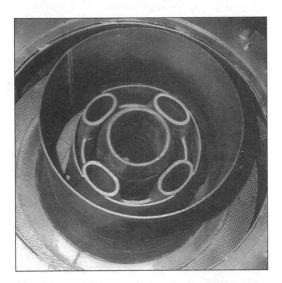

Figure 6.7 The face of a typical low-NO$_x$ burner showing the ducts feeding the pf/air and air streams.

Figure 6.8 Typical low-NO$_x$ burners firing in a power station combustion chamber.

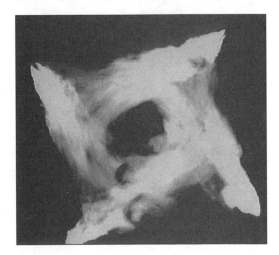

Figure 6.9 Tangential firing in a combustion chamber.

9 of nitrogen to oxygen in air. If combustion is with pure oxygen, then 3.76 is replaced by 0.00, and for oxygen-enriched air z takes the appropriate intermediate value. For stoichiometric combustion $z = 0$. Thus, a fuel whose composition is known on a weight basis rather than a molar basis, such as coal containing, for example, 84% by weight of carbon, 12% by weight of hydrogen, 1.5% by weight of oxygen, and 2.5% by weight of sulfur would undergo the following overall stoichiometric reaction:

$$10.031O_2 + 37.7N_2 + C_7 H_{12} \, C_7H_{12}O_{0.0938}S_{0.0781}$$

$$= 7CO_2 + 6H_2O + 0.0781SO_2 + 37.7N_2$$

Generally, coal is burned with an excess amount of air (called excess air) typically in the range 5–25%, so that smoke formation, etc., is minimized and this air would be additional to the quantity given above. A detailed description of the method of calculation of flue gas compositions is given in Appendix 3.

The departure from stoichiometric proportions also can be described by the use of the equivalence ratio ϕ. This is defined as the ratio of fuel available to the amount required for stoichiometric combustion, i.e., [fuel/air] $_{actual}$/[fuel/air] $_{stoich}$. Lean mixtures have $\phi < 1$ and rich mixtures have $\phi > 1$.

Two other definitions are often used in boiler practice:

1. The CO_2 efficiency η_{CO_2} defined as

$$\eta_{CO_2} = \frac{\text{actual mol fraction of } CO_2 \text{ produced}}{\text{mol fraction of } CO_2 \text{ produced by complete combustion}}$$

2. The % carbon burn-out η_C defined as

$$\eta_C = \frac{\text{\% carbon burned}}{\text{\% carbon in original fuel}}$$

This second method is useful for high-ash-content fuels or coal-containing liquid fuels, but care has to be taken in correcting for the effect of ash in the fuel. Effectively, the method involves measuring the carbon content in the stack solids by burning off the carbon from a weighed sample of stack solids and then estimating the total carbon emissions rate in the flue gases for a known firing rate. There are various important definitions in relation to coal combustion, namely, loss on ignition (LOI) at $100 - \eta_C$, and carbon in ash, which is measured as described in the preceding paragraph.

Combustion occurs via free radical reaction with overshoot in the concentrations of H, O, and OH in the flame zone. These recombine, but at flame temperatures the combustion products are dissociated and can be represented by the dissociation reactions

$$CO_2 \xrightarrow{\;\;1\;\;} CO + \tfrac{1}{2}O_2 \qquad\qquad \text{R 6.1}$$

$$H_2O \xrightarrow{\;\;2\;\;} H_2 + \tfrac{1}{2}O_2 \qquad\qquad \text{R 6.2}$$

$$H_2O \xrightarrow{\;\;3\;\;} \tfrac{1}{2}H_2 + OH \qquad\qquad \text{R 6.3}$$

$$\tfrac{1}{2}H_2 \xrightarrow{\;\;4\;\;} H \qquad\qquad\qquad \text{R 6.4}$$

$$\tfrac{1}{2}O_2 \xrightarrow{\;\;5\;\;} O \qquad\qquad\qquad \text{R 6.5}$$

together with

$$\tfrac{1}{2}N_2 + O_2 \xrightarrow{\;\;6\;\;} NO \qquad\qquad \text{R 6.6}$$

Consequently, the products at the flame temperatures contain small quantities of carbon monoxide, hydrogen, etc., and the amounts of these products may be obtained by computation by means of standard techniques, e.g., Kuo (1986) and a method for dissociated combustion products is outlined in Appendices 4 and 5. Numerous commercial computer programs are available to undertake these calculations. NO does not reach its equilibrium concentration in the time available in most coal-fired combustion chambers, although many other species attain equilibrium levels. The concentration of metal compounds, e.g., in slags, can also be calculated.

Flue gas composition can readily be measured by withdrawing samples and, after filtration to remove ash, etc., is analyzed for CO, CO_2, and unburned hydrocarbons by infrared methods and for O_2 by paramagnetic instruments or zirconia probes. Generally, measurements of O_2 and/or CO_2 are made during the installation and setting up of smaller boilers, while larger plants will have instrumentation installed for continuous monitoring of the plant. NO_x can be measured by chemiluminescent methods, giving both NO and NO_2, or by UV methods that normally only give total NO_x. SO_2 and SO_3 can be measured by infrared methods. In all cases care has to be used if samples are dried, and the best results are obtained by using heated lines and heated measuring instruments; results are often normalized to a standard amount of excess air (often 3% O_2, sometimes zero) or percent CO_2. Smoke can be measured across the flue duct by light beam methods, but this gives no indication of the amount of the larger nonlightscattering cenospheres. Total stack solids can be measured only by using isokinetic sampling to collect both gas-phase soot and cenospheres on a filter (often sintered bronze or fiberglass) and determining gravimetrically the amount for a known volume of gas sampled. VOCs and PAH can be determined in the gas phase by absorption techniques or from solids (e.g., soot) by extracting the solid sample with an organic solvent and analyzing by HPLC. Metals can be determined by the usual techniques from the collected solid samples.

Measurements within combustion chambers have to be made with water-cooled probes, and the solid sample is normally captured by a sintered metal filter cone situated within the probe. Gases are analyzed as described above. Minor in-flame species such as HCN and NH_3 can be determined by dissolving the samples in water and analyzing the solution by ion probes; care may have to be taken to prevent interference effects.

Theoretical adiabatic flame temperatures may be derived from the heat released by the combustion process, but the enthalpy change associated with the combustion process must take into account the influence of dissociation. An outline of one method of computation is given in Appendix 5. Calculations of this type have shown that the maximum theoretical flame temperatures for the combustion of pf with air are 2100K for zero excess air (stoichiometric mixture), 1900K for 10% excess air, and 1800K for 20% excess air. Coal flames can have slightly lower flame temperatures since it is dependent upon the ash and sulfur contents. In practical flames, heat losses to the furnace and in the waste gases greatly reduce the flame temperatures actually attained. These temperatures can be obtained only by experimental techniques or by heat balance calculations, and as a consequence, practical flame temperatures are generally in the region of 1400–1900K.

Flame temperatures can be calculated or experimentally determined by suction pyrometers or by shielded thermocouples, and equipment for these techniques are widely available commercially (e.g., Land, International Flame Research Foundation). In research studies of industrial combustors, temperatures can be measured by sodium D-line reversal or infrared absorption-emission techniques.

Flue stack losses can be calculated or can be estimated graphically as indicated in the Appendix. Many instruments can measure both flue-gas temperatures and CO_2 content and can compute combustion efficiency.

Combustion intensities   The combustion intensity I for any combustion system is given by

$$I = \frac{FH_c}{V_c p} \quad \text{W/m}^3/\text{s/bar} \qquad \text{E 6.1}$$

where F is the fuel feed or firing rate, H_c is the enthalpy of combustion, V_c is the chamber volume, and p is the pressure, which is normally 1 bar except for power plants operated at high altitudes. Obviously, the smaller the combustion volume for a given fuel-feed rate the higher the combustion intensity. The rate of energy release is virtually totally controlled by the coal-particle burning; any burn-out of residual carbon or carbon monoxide first formed makes only a small contribution to the total heat release. The combustion intensity may be calculated on the following basis.

If the air quantity is Q, then at stoichiometric firing the air flow rate is FQ; thus with $E\%$ excess air, the total rate is $FQ\,(I+E/100)$. If there is little change in volume due to the number of moles of the combustion products, then the hot-gas flow rate is given by

$$\frac{FQ(1+E/100)}{p}\left(\frac{T_f}{T_0}\right) \quad \text{m}^3/\text{s/bar}$$

where T_f and T_0 are the flame and initial temperatures, respectively. This is the modified Rosin equation (Essenhigh, 1967). Thus the chamber volume V_c is given by

$$V_c \frac{FQ(1+E/100)}{p}\left(\frac{T_f}{T_0}\right)t_B$$

where t_B is the burning time. Thus, we can get

$$I = \frac{H_c}{Q(1+E/100)(T_f/T_b)t_B} \quad \text{W/m}^3/\text{bar} \qquad \text{E 6.2}$$

Generally, H_c/Q is approximately constant, so the combustion intensity depends primarily on the burning time t_B. The burning time varies linearly with particle diameter. If the equation is applied to the case of pf combustion with an initial particle diameter of 100 μm and a burning time of 1–2 s, this would imply a combustion intensity of about 0.5 MW/m^3/hr/bar. Combustion intensities are higher in very turbulent systems or where preheated air is used, and values are about 5MW/m^3/hr/bar. Combustion intensities on a grate, however, would be about 0.1 MW/m^3/hr/bar (Essenhigh, 1967). Generally, spray combustion has higher combustion intensities than coal by approximately an order of magnitude, while the combustion of gaseous fuels is two orders of magnitude higher.

Combustion chamber performance   The performance of a combustion chamber, whether in a furnace or a boiler, can be modeled by the zone method or by means of a full computational fluid flow approach. In the case of coal-fired furnaces the presence of the coal particle and ash is an important aspect of the calculation procedure, that is, it is a two-phase system.

The computational fluid flow approach can give detailed information on flow and heat transfer to specific parts of the furnace, e.g., to individual boiler tubes and the formation of pollutants; but its accuracy, at least at the present time, is still not very high.

Ash and Slag Deposition

Ash deposition is one of the most important operational problems associated with the efficient utilization of coal (IEA Coal Industry Advisory Board, 1995). Since deep cleaning of coal is expensive (Couch, 1994), ash is present in all coal-fired furnaces and must be carefully controlled.

Equipment manufacturers have used several approaches for ash management to accommodate effective collection and disposal of the deposit. Dry and wet bottom furnaces utilize very different operational conditions to achieve this goal. Most pulverized coal units that are offered today are of the dry bottom type, although wet bottom or slagging bottom furnaces may still be offered for special applications.

Since the presence of ash is unavoidable, coal-fired power stations are designed to tolerate some deposit on tube surfaces without undue interference of unit operation. Knowledge of ash deposition tendencies of coals is important for boiler manufacturers since boiler design features can be varied to accommodate "difficult" coals. Some manufacturers accommodate various ash characteristics by adjustment of furnace dimensions and the number of deposition removal systems such as wall blowers. The criteria of several utility boiler manufacturers for designing boilers to avoid deposition have been discussed by Barret (1990). It was observed that each manufacturer applied a different set of criteria and placed different emphasis on the coal analyses details used for prediction of ash depositional behavior.

The occurrence of extensive ash deposits can create the following problems in a boiler: (1) reduced heat transfer due to a reduction in boiler surface absorptivity and thermal resistance of the deposit, and (2) impedance of gas flow due to partial blockage of the gas path in the convective section of the boiler.

The coal-fired boilers are operated mainly under conditions where the ash is removed as a dry particulate solid. In such cases there is considerable industrial interest in predicting the properties of ash produced and the tendency to form slags. This is entirely a question of the composition of the main inorganic constituents in the coal, but the problem of relating this composition to actual performance in practice is difficult. A number of correlation equations have been produced (e.g., Badin, 1984, Couch, 1994) to try to predict these effects.

Some boilers and furnaces (and gasifiers) are designed for operation at higher temperatures and the ash is allowed to melt and flows out as a slag. This slag gives a glassy substance when solidified that can be used as a building material and can have environmental advantages if used for landfill.

As far as deposits are concerned, these are caused by chemicophysical reactions. Five principal types of deposits are known:

1. Slag deposits, either loosely held or strongly bonded on radiatively heated furnace walls and other surfaces
2. High-temperature fouling deposits, bonded to convection-heated surfaces such as superheaters and reheaters
3. Corrosion deposits, a subgroup of high-temperature fouling deposits
4. Low-temperature fouling deposits (acid smuts), on cooled surfaces such as air heaters and economizers near the outlet point of the unit
5. Low-temperature slag deposits, tenaciously bonded deposits that may form in the same region as the low-temperature fouling deposits, although on surfaces that are somewhat hotter than those for the low-temperature fouling deposits

Generally, the prediction indices are based on the inorganic species that are divided into groups such as the base (B) group Fe_2O_3 + CaO + MgO + Na_2O, the acid group, SiO_2 + Al_2O_3 + TiO_2, and the ratio B/A used as an ash descriptor (Carpenter and Skorupska, 1993). Other parameters exist for ash viscosity, slagging propensity, and fouling propensity.

Advanced models can give information relating to the fouling occurring at low temperatures or a high-temperature index related to deposition in the superheater or reheater.

6.6
Fluidized-Bed Combustion

Fluidized-bed combustion differs from both conventional stokers and pulverized-fuel combustion in being applicable to a wide spectrum of coal-fired systems ranging from small industrial boilers and furnaces to large power generation units. The diverse uses of fluidized-bed combustion have given rise to a number of different types of combustor design. Fluidized-bed combustion may be used for furnace applications, for example, to produce a hot gas for drying purposes. In order to reduce the gas temperature drying, high excess air levels or flue gas recirculation are often employed. The main categories are described next.

Atmospheric Pressure FBC Boilers

In boiler applications, heat transfer surfaces such as tubes may be placed in the fluidized bed to raise steam or to produce hot water. Further heat may be recovered from the combustion gases using convective heat exchangers as shown in Figure 6.10.

As a result of the turbulent nature of fluidization, the heat transfer coefficients between the bed material and the immersed tubes are generally high in fluidized beds as compared with heat transfer coefficients in convective heat exchangers. The bed temperature is maintained in the range 750–950°C and is therefore considerably lower than the operating temperatures of other coal-fired combustion systems. The bed consists of coal ash, sand, and limestone that is added to control SO_2 emissions.

The combustion gases leave the bed, pass through the freeboard, and exit near the top. As bubbles burst near the bed surface, sprays of bed material are thrown into the freeboard. The purpose of the freeboard is to allow sufficient space above the bed for these particles to fall back to the bed.

The gases leave the combustor at approximately the bed temperature, and the exact value depends on the degree of combustion and heat transfer in the freeboard. The gases then pass into a convective section where further heat is recovered and the gases are cooled to the required stack exit temperature (usually 150–200°C).

The total heat release rate is of the order of 1 MW (thermal) per m^2 of bed surface at a fluidizing velocity of 1 m/s and a pressure of 1 bar.

The operating temperature of a fluidized-bed combustor usually lies in the range 750–950°C. The upper limit is determined by the need to avoid ash or bed material sintering. The bed temperature also has an important effect on the sulfur retention efficiency. At low bed temperatures, sulfur retention is inhibited by poor calcination of the added limestone and at high bed temperatures by decomposition of the calcium sulfate. As a result there is an optimum temperature for sulfur capture at about 850°C. If, therefore, sulfur retention is required and

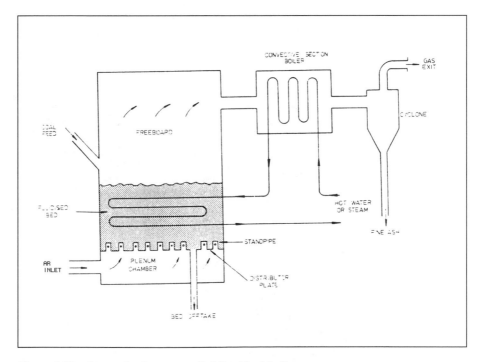

Figure 6.10   Atmospheric pressure fluidized-bed boiler.

the fuel has a high sulfur content, the retention efficiency is usually the main consideration determining the bed temperature.

Operation at a high excess air level increases the mass flow rate of hot stack gases emitted to the atmosphere and consequently reduces the boiler efficiency as shown in Figure 6.11. This efficiency loss increases linearly with the excess air level. Operation at a low excess air level, however, leads to a high loss of combustible material that is mainly unburned char entrained in the gas stream together with some combustible gases.

The above factors result in the existence of an optimum excess air level for maximum boiler efficiency as shown in Figure 6.11. The position of the optimum varies with the other combustion design parameters, for example, bed temperature, bed depth, and fluidizing velocity, but usually lies in the range 10–40% excess air. For a system fluidized by air (without preheat), all the heat

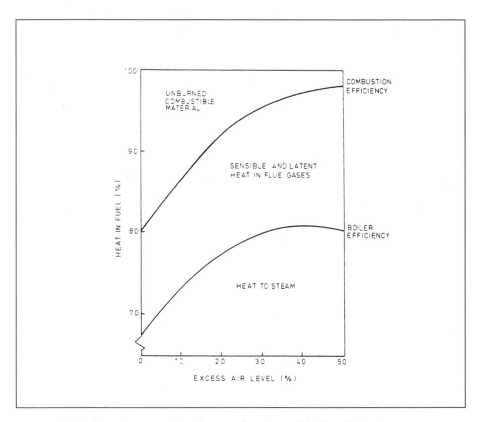

Figure 6.11   Effect of excess air level on combustion and boiler efficiencies.

would be removed by combustion gases if about 150% excess air were used. At an excess air level of 30%, approximately half of the heat released is therefore removed by the combustion gases and half by heat transfer from the bed. In industrial boilers, an excess air level of 30–50% would normally be used, giving a boiler efficiency of about 80%. For power plant applications, however, high-efficiency designs would be favored and an excess air level of 15% would be typical. Two other important, and related, parameters are the calcium acceptor/sulfur ratio and the bed depth, discussed below. Another design factor that can be introduced is air staging to reduce NO_x formation; here, some of the air is added into the freeboard.

Calcium/sulfur ratio   The proportion of the sulfur in the coal that can be retained as calcium sulfate by the addition of an acceptor to the bed depends on the feed rate of the acceptor. The feed rate is usually expressed as the calcium to sulfur mole ratio; a value of 1 corresponds to the stoichiometric calcium feed rate. For limestone, the stoichiometric feed rate is 3.1% of the coal feed rate for every 1% of sulfur in the coal. To retain 85% of the sulfur, for example, would typically require a calcium to sulfur mole ratio of 2/4, with the exact value depending on the other design parameters. The reactivity of the limestone is also extremely important and can in itself require a change in the calcium to sulfur mole ratio of a factor of 2. However, in some cases the required SO_2 emission can be achieved by adding only a small amount of limestone.

Bed depth   The bed depth for fluidized-bed boilers is of the order of 1 m if full sulfur retention is required, but otherwise 0.15–0.5 m is usually employed. The minimum bed depth that may be used in a design is determined by two factors. First, there must be sufficient free space in the bed for lateral mixing to occur in order to avoid temperature gradients, and second, the bed has to accommodate the tube bundle (if included) without the tubes being too close together.

Although a bed depth greater than the minimum may be used, this adds to the pressure drop and therefore to the fan power requirements. Since fan power is an important operating cost consideration in most boilers, it is usual to design for the minimum bed depth. As the bed depth is reduced, however, the combustion and sulfur retention efficiencies decrease and the amount of combustion in the freeboard increases. In some cases, it is therefore necessary to design for more than the minimum bed depth specified above in order to achieve the required efficiency targets.

The mean particle diameter of the bed material is usually in the range of 0.5–1.5 mm, with fluidizing velocities in the range 1–3m/s, respectively. The maximum particle size is in the range of 2–6 mm. A high fluidizing velocity requires a coarser bed in order to obtain good fluidizing conditions and to avoid excessive carryover of bed material.

A high fluidizing velocity also leads to more combustion in the freeboard and increases the carryover of unburned char and unconverted sulfur acceptor, thereby reducing the combustion and sulfur retention efficiencies. If the reduction in performance is unacceptably large, design changes such as refiring the elutriated material may become necessary. However, the logical extension to this is to use a circulating fluidized bed, described in the next section.

Circulating Fluidized-Bed Systems

The problem with conventional fluidized-bed combustion is that as combustion proceeds, the ash level builds up, and when this is removed a certain amount of unburned char is removed with it. The circulating bed has been devised, in part, to overcome this problem. Circulating fluidized beds operate with relatively high gas velocities and fine particle sizes, and involve a different fluidizing regime from that in conventional fluidized beds. Because the maintenance of steady-state conditions in a fast fluidized bed requires the continuous recycle of particles removed by the gas stream, the system is referred to as a "circulating" bed. However, the term "circulating" bed is often used to include fluidized-bed systems containing multiple conventional bubbling beds between which bed material is exchanged.

A typical fast fluidized-bed combustion boiler is illustrated in Figure 6.12. The main combustion and sulfur-retention reactions occur in a highly turbulent, nonbubbling fluidized bed usually at least 10 m deep. The fluidization conditions in the "fast" fluidized-bed reactor are created by a combination of a high fluidizing velocity (usually about 10 m/s) and fine bed material (mean particle size about 150 μm). In such a system, the solids rapidly become entrained in the gas flow and are removed from the containment vessel. Recycle of the material is therefore essential to maintain steady-state conditions, the rate of recycle determining the solids concentration in the reactor. The solids-laden gas stream leaving the fast bed passes first to a separation device, usually a cyclone. The solids recovered are then fed to a conventional fluidized-bed reactor containing a tube bundle. This bed extracts heat from the recycling solids to heat water or

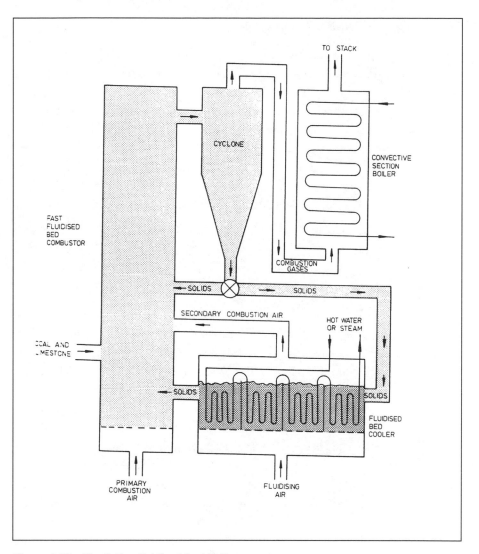

Figure 6.12   Circulating fluidized-bed boiler.

to raise steam, usually in a series of stages. The solids leave the bed cooled, typically to about 100°C and are then reinjected into the fast fluidized-bed combustor. The recycling solids may therefore be regarded as acting as a heat carrier to remove the heat generated in the combustor, which generally operates at a low excess air level. The cooler is fluidized by air at a low velocity (usually less than 1 m/s).

Pressurized Fluidized-Bed Combustion

At present, gas turbines cannot be operated by direct coal firing, and combined cycle systems are fired by distillate fuel oil or a gas. In spite of considerable research and development in the 1950s and early 1960s, it has not proved possible to operate a gas turbine directly on coal using pulverized-fuel combustion. The difficulty lies in obtaining an adequate lifetime for the gas turbine blades when expanding the gas produced by coal firing. The way around this is to use exhaust gases directly from a pressurized bed combustion, and the efficiencies obtained are indicated in Figure 6.13. The technology is in competition with IGCC (see Chapter 7) and is dependent on hot gas filtration of the gases so that the gas turbine sees only an acceptable loading and acceptable particle sizes.

6.7
Scaling Criteria for Burners and Furnaces

There has been considerable research undertaken over a number of years to develop reliable scaling laws for the scale-up of industrial combustors. When a burner is scaled the ultimate aim is ideally to achieve similarity in all fluid dynamic and thermochemical-related processes in the scaled flame. If this is achieved, the performance of the scaled burner is identical with the full-scale situation. In reality, compromises are necessary because not all the physical and chemical processes scale in the same way. Providing these phenomena of interest can be described using a set of equations, then by nondimensionalizing the equations, all the required data can be obtained. However, if all the required equations are not available, then the assumption must be made regarding the type of processes at work in order to obtain the required parameters from the nondimensional groups. This process will lead to significant uncertainty in the scaling laws.

This technique has not been used much with bed combustion or fluidized bed, and most effort has been devoted to pf combustion. The most important scaling considerations with respect to flames of pulverized coal are related to the effect of scale on the gas-phase fluid dynamics (turbulent mixing), and two-phase interactions between coal particles and the main flow, and the consequential effects on the resultant in-flame thermochemistry.

These criteria have been set out by Smart et al. (1998, 1997) and Weber (1996). Traditionally, there are two practical criteria for the scaling-down of

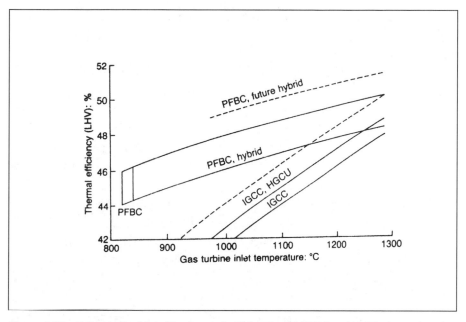

Figure 6.13 Estimated achievable thermal efficiencies for PFBC and IGCC plants: band that represents PFBC hybrid technology shows possible efficiency increases when gas turbine inlet temperature is increased; band relating to IGCC plants shows recent estimates of achievable efficiencies; dotted lines refer to higher efficiencies for future PFBC and IGCC plants; condenser pressure is 0.002 bar (Pillai, 1992).

swirl-stabilized pulverized-coal burners: constant-velocity and constant-residence (constant mixing time) scaling, as discussed previously in the papers by Smart et al. and Weber et al. Both of these scaling criteria rely on the scaling of the large macroscale turbulent-mixing process. Geometric similarity of the burner is maintained in both cases, and swirl number is maintained constant. Constant-velocity scaling is the most commonly used by burner manufacturers; in this case, velocity and momentum ratios of coaxial combustion-air and fuel injection are maintained constant with scale reduction. For constant-residence-time scaling, the burner velocities scale with the burner diameter as scale is reduced, but coaxial velocity and momentum ratios are still maintained constant.

Scaling of NO_x Emissions

In the previous section it was shown that a partial modeling approach does not follow hard and fast rules, and the process condition has a significant influence on choosing the scaling parameters. In recent years the application of dimensionless analysis on NO_x formation has received considerable interest (e.g., Chen and Discoll, 1991).

Extensive fundamental studies have been carried out to investigate the effects of initial velocity, burner diameter, and fuel composition on NO_x-emission scaling. However, the majority of these investigations include fuels of simple chemical composition. Unlike the pragmatically proved scaling law, which provides design guidance for practical appliances, fundamental investigation provides a foundation for the understanding of the complex fluid mechanics and chemical process, to properly scale NO_x emission in industrial burners.

NO_x scaling is controlled by a number of linked factors such as fuel composition, the controlling pathway on NO_x formation mechanism, radiation properties of the flame, influence of flame strain on NO_x formation, etc. Taking into account the complexity of such a process, the idea of deviation of a leading-order or simple scaling law has many attractions. However, the inaccuracy of the simple scaling law has led to extensive fundamental research into the development of a leading order scaling law. In order to investigate the NO_x scaling, the emission index EI_{NO_x}, which is a dimensionless quantity, is used for convenience. It is defined as the ratio of mass of NO to the mass of fuel burned by the combustion process, namely,

$$EI_{NOx} = \frac{m_{NO\ emitted}}{m_{F\ burned}} \qquad\qquad E\ 6.3$$

EI_{NO} for hydrocarbon fuel combustion can be obtained from concentration measurements of the NO_x and all of the carbon-containing species, i.e.,

$$EI_{NO} = \left(\frac{X_{i\ NO}}{X_{i\ CO} + X_{CO_2}} \right)\left(\frac{x\ MW_{NO}}{MW_F} \right) \qquad\qquad E\ 6.4$$

where x is the number of carbon atoms in a mole of fuel, MW is the molecular weight of NO and of fuel, respectively, and X_i is the mole fraction of species i.

A simple leading-order scaling law for EI_{NO} can be obtained for a simple burner firing natural gas fuel by relating it to the flame volume, i.e., the diameter d and initial velocity U_0 of the burner:

$$EI_{NO_x} = \left(\frac{d_0}{U_0} \right) const. \qquad\qquad E\ 6.5$$

The above zeroth-order scaling will be valid if geometrical, thermal, and chemical similarity is kept. The extensive research on simple NO_x scaling indicates many failures in this approach, and a number of factors responsible for these failures have been envisaged.

However, recent studies by different investigators have provided valuable insight into the combined role of a nondimensional group on NO_x scaling. It is of prime importance to indicate that most of the experimental and theoretical studies have focused on thermal-NO formation and prompt and the fuel-NO formation pathway have not been studied in great detail. Recently, two jet flame models have been developed that specifically include prompt-NO in the scale-law formulation: the two-zone asymptotic model of Rokke et al. (1992) for gaseous fuel/air, that has been extended for oxygen enriched and preheat gaseous fuel combustion, has been validated against a considerable database which now exists for hydrocarbon flames. In this model a simplified mechanism for the combined thermal-, prompt-, and nitrous oxide-NO formation pathway is analyzed.

Based on this approach, a leading-order scaling law, taking into account prompt-NO, has been developed, which leads to the equations:

$$EI_{NO_x} = 22 - Fr^{0.4}Y_F^{-0.2}\left(\frac{d}{U_0\rho_F}\right)\left(1.65X_f^{-0.9} - 0.35X_f^{-0.4}\right) \qquad \text{E 6.6}$$

$$EI_{NO_x}\frac{\rho_0 U_0}{d_0} = 57.3e^{-4.5\Delta X_{O_2}}\left[5\left(\Delta X_{O_2}^{1.15} + 4.71\times10^{-2}\right)X_f^{7/8}\right.$$

$$\left. + e^{1.5}\left(-X_f^{0.45} + 10\Delta X_{O_2}^{0.7} + 5.33\times10^{-4}\Delta T_{f_p}\right)\right]\left(\frac{\gamma f_p}{\gamma}\right)F_\gamma^{3/5} \qquad \text{E 6.7}$$

where the terms are defined by Smart et al. (1997, 1998) and Yap et al. (1998).

It is feasible to extend the leading-order scaling law to include the effect of fuel-NO on the emission index. The NO formation during coal combustion is a complex process when the fuel-NO pathway is a dominant route. These models permit the estimation of a volatile NO_x and the char-NO_x as well as the thermal-NO_x. The detailed investigation indicates that the NO formation process is indeed complex, and in order to have a robust leading-order NO_x scaling law, the effect of nonequilibrium chemistry, fuel- and prompt- NO effects, flame radiation, coupling of chemistry, and radiation with the turbulent flow field and fuel-NO formed from heterogeneous sources should be considered.

Extensive research in the coal area has shown the effect of fuel-N on total

NO_x emission. This research has indicated the relative split between the fraction of NO emissions resulting from thermal-NO_x, volatile-NO_x, and char-NO_x components. Data clearly show that about 20% of the total NO_x was thermal-NO_x when burning coal. Of the balance, the volatile nitrogen was shown to be the primary source of the fuel-NO_x during normal firing, with a strong dependency on the local air/fuel ratio. However, char-NO_x was found to be relatively independent of oxygen concentration or burner geometry. Based on these results, it may be concluded that scaling of the coal burner will not affect the char-NO_x emission; however, it will have a significant influence on the volatile-NO_x and thermal-NO_x.

It has been argued that a simple form of scaling involves the assumption that the concentration of NO_x formed in a drop tube is equal to that in a power station. This is true only if the route to NO_x formation is from the char.

6.8
Computational Fluid Dynamics Methods

Computational fluid dynamics (CFD) computational methods applied to combustion have made very considerable advances over the last few years. In most respects they have replaced a number of simple coal combustion modeling techniques developed over a number of years, such as plug flow and well-stirred reactor assumption (Smith et al., 1981, Abbas et al., 1996). The advent of powerful computers means that CFD methods can be applied at levels of sophistication ranging from personal computers to very large parallel computers.

Combustion

Methods One of the main features of using finite volume methods is that additional equations for the enthalpy, species concentrations, etc., can be included and the method still converges. For gaseous combustion the additional equations can be written in the same format as the flow equations, but thermal radiation has to be treated separately. For coal-particle and oil-droplet combustion, a Lagrangian droplet or particle tracking method is usually preferred (Abbas et al., 1996).

A modern coal combustion CFD code for combustion typically contains the following models:

- For flows: equation for enthalpy, equation of state, continuity, and for the Navier-Stokes equations together with turbulence models, k-ε, ASM, and RSM.

- For gaseous combustion of the volatiles or added gases: models are based on eddy breakup, "mixed-is-burned" assumption, mixture fraction (f) and variance (g), assumed pdf (clipped Gaussian, β function, etc.), combustion model, involving flamelets or similar assumption, and /soot or other pollutants model (Bray, 1996).

- For coal combustion it is necessary to invoke much more complex models, including flow of the discrete particle, devolatilization (single, double, or multiple competing reactions), char oxidation, volatile combustion model, formation of NO_x/soot model, and an ash slagging model (Jones et al., 1998).

- For thermal radiation: discrete transfer and/or Monte Carlo model.

These models require much more physical property data and other input data than do heat balance models.

In many gas flames it is sufficient to describe the combustion by the "mixed-is-burned" model. In this, the fuel and oxidant are assumed to react instantaneously, producing products:

$$1 \text{ kg fuel} + i \text{ kg oxidant} = (1 + i) \text{ kg products}$$

where i is the stoichiometric ratio. Thus, in the furnace there are only regions of fuel and products or of oxidant and products. This has the great advantage that the equations for the mass fractions of fuel and oxidant can be combined into a single equation for the mixture fraction f. Furthermore, by specifying a probability distribution function (pdf), allowance can be made for the fact that the mass fractions fluctuate: at any point at one instant the mixture may be fuel and products, and at a later time oxidant and products. Several choices have been made for the pdf, but they usually depend on f and its variance g, both of which are calculated from equations having the same structure as all the others, and of course, additional constants. In coal flames there is an initial devolatilization, and a remnant char is formed that burns more slowly and leaves inert ash.

Several different methods have been used for the radiative heat transfer. In TEACH and other early CFD codes, the discrete ordinates model was favored. This had the advantage that it could be discretized by finite differences, but it could not be extended easily to complex three-dimensional geometries. Most CFD codes now use the Discrete Transfer Model, which is a simplified version of Monte Carlo. Both Discrete Transfer and Monte Carlo are based on tracking photons through the combustion domain. This has the advantage that absorption by the intervening gas and scattering by particles can easily be included. Both methods are, however, very different from all the other finite volume equations and, because of computing time, are usually solved on a coarser grid than the flow and other equations.

Attention has to be paid to pf burners; that is the area where research and development is currently directed and in these to convergence obtaining grid-independent solutions. These points are obvious but nowadays often ignored, especially when much effort has been expended on creating a complex three-dimensional grid.

Profiles of the velocity, the turbulence parameters k and ε have to be specified at the inlet to the turbulent flow domain. Fully developed profiles may be used when there is a long inlet pipe or channel, but unfortunately, this is not the case in most combustion applications, such as wind boxes and burner mouths. Also, the flows are often swirled, so the swirl as well as the axial velocity is required. "Rules of thumb" have been developed, but these should be checked for each new application and their importance assessed.

There are more advanced turbulence models, termed ASM and RSM, but unfortunately, these have not produced the improvements expected, the results depending very much on the flow of interest. Thus the k-ε model, with all its known faults, is still used in most engineering calculations.

In view of the above, turbulent flow modeling still requires much expertise and many different runs, if the results are to be credible.

Turbulent Combustion Modeling

Among the various problems in CFD modeling (e.g., spatial discretization, turbulence modeling, chemical kinetics, etc.) the problem of turbulent combustion deserves to receive much attention. The turbulent combustion model is defined here as the model that describes the general case of statistically three-dimensional time-dependent flow and that predicts at a minimum the mean velocity and composition field.

Turbulent combustion occurs when the chemical and thermal processes of combustion interact with the turbulent flows. This interaction occurs because of close similarity of the various time scales associated with these processes. In turbulent combustion, the fluctuations in the temperature and concentration fields are as important as in the velocity field. These fluctuations, either in the initial inert mixture or as a result of combustion, will affect the flow structure. From the point of view of CFD, the turbulent combustion effects are simply introduced into computation of the turbulent fluxes and, mainly, of the volumetric combustion rate for any interesting species.

The main areas of turbulent combustion modeling of diffusion gaseous reactants could be summarized as follows:

1. Analytical approaches that treat simplified models of the governing equations: mechanism, improved numerical discretization scheme, and liner equation solvers.

2. Engineering calculation method: This type of turbulence modeling has received the most interest. This is perhaps because of its direct relevance to the design and development stages of industrial burners. The interest in this area comes from the fact that it can be performed using the present computational capabilities while producing qualitatively acceptable results. This is the area where the CFD technology can play a major role now and in the future.

3. Numerical simulations that employ detailed modeling of turbulent reacting flows: The application of direct numerical simulation of turbulence phenomena to simple reacting flows is just starting recently to emerge. The large eddy simulation (LES) and the direct numerical simulation (DNS) in which the Navier-Stokes equations are solved directly without a turbulence model have been lately used to obtain instantaneous flow condition for nonreacting cases. Although these methods will become a more useful technology in the future, they have been applied to a very few cases of low-Reynolds-number reacting flows and require huge computational resources.

4. Studies-oriented practical applications, e.g., those concerned with reduced reaction: This represents an important weight in applying the developed theoretical models into practical CFD codes, because the success of the physical submodels in the computational algorithm is restricted by other numerical and computational factors (e.g., grid adaptation, boundary-conforming discretization, and convergence of the linear equation solvers). Therefore, these aspects represent important research area for improving the performance of the computational codes.

Governing Equations of Turbulent Combustion

Since reacting flows are of interest here, Favre or density-weighted averaging is used for all dependent variables except pressure and density. Pressure P and density ρ are decomposed into the conventional (time, space, or ensemble, as appropriate) average and fluctuations. Due to the strong fluctuations of the density that occurs in reacting flows, it is more common to use the density-weighted values rather than the conventional mean. Here and below, overbars ($^{-}$) indicate conventional averaging, while the tilde (~) indicates density-weighted or Favre mean values. The instantaneous density and pressure could be decomposed as

$$\rho = \bar{\rho} + \rho' \quad P = \bar{P} + P'$$

E 6.8

while the other variables are decomposed as

$$\phi = \tilde{\phi} + \phi''$$

E 6.9

where the Favre mean value is defined as

$$\tilde{\phi} \equiv \frac{\overline{\rho\phi}}{\bar{\rho}}$$

E 6.10

The fluctuating components have the following properties:

$$\bar{\phi}'' = -\frac{\overline{\rho'\phi''}}{\bar{\rho}} \quad \overline{\rho\phi''} = 0 \text{ and } \bar{\phi}' = 0$$

E 6.11

The relationship between the Favre average and the conventional average is

$$\bar{\phi} = \tilde{\phi} + \bar{\phi}''$$

E 6.12

Thus, one kind of averaging can be related to the other.

The Favre averaged equations that govern the statistically stationary turbulent reacting flowfield under the influence of gravity may be written as

$$\frac{\partial}{\partial x_i}\left(\bar{\rho}\tilde{u}_i\right) = 0$$

E 6.13

$$\frac{\partial}{\partial x_j}\left(\bar{\rho}\tilde{u}_i\tilde{u}_j\right) = -\frac{\partial \bar{p}}{\partial x_i} - \frac{\partial}{\partial x_j}\left(\bar{\rho}u_i''u_j''\right) + \bar{\rho}g_i$$

E 6.14

In the above equations, u is the velocity component in direction i and x is the position vector. Similarly, the transport equation for the mean species concentration could be expressed as

$$\frac{\partial}{\partial x_j}\left(\overline{\rho}\tilde{u}_j\tilde{Y}_a\right) = -\frac{\partial}{\partial x_j}\left(\overline{\rho u_i'' Y_a''}\right) + \overline{\rho}\tilde{S}(Y_a) \qquad \text{E 6.15}$$

The above equations contain turbulent momentum and scalar fluxes that are approximated via the two-equation k-ε turbulence model

$$\frac{\partial}{\partial x_j}\left(\overline{\rho}\tilde{u}_j k\right) = \frac{\partial}{\partial x_j}\left(\frac{\mu_t}{\sigma_k}\frac{\partial k}{\partial x_j}\right) - \overline{\rho u_i'' u_j''}\left(\frac{\partial \tilde{u}_i}{\partial x_j}\right) - \overline{\rho}\varepsilon \qquad \text{E 6.16}$$

$$\frac{\partial}{\partial x_j}\left(\overline{\rho}\tilde{u}_j\varepsilon\right) = \frac{\partial}{\partial x_j}\left(\frac{\mu_t}{\sigma_\varepsilon}\frac{\partial \varepsilon}{\partial x_j}\right) - C_{\varepsilon1}\frac{\varepsilon}{k}\overline{\rho u_i'' u_j''}\left(\frac{\partial \tilde{u}_i}{\partial x_j}\right) - C_{\varepsilon2}\overline{\rho}\frac{\varepsilon^2}{k} \qquad \text{E 6.17}$$

Hence, the Reynolds stress is specified as linearly related to the mean rate of strain via a scalar turbulent viscosity

$$\overline{\rho u_i'' u_j''} = \frac{2}{3}\delta_{ij}\left(\overline{\rho}k + \mu_t\frac{\partial \tilde{u}_i}{\partial x_i}\right) - \mu_t\left(\frac{\partial \tilde{u}_i}{\partial x_j} + \frac{\partial \tilde{u}_j}{\partial x_i}\right) \qquad \text{E 6.18}$$

and the turbulent scalar flux obtained using a gradient diffusion approximation

$$\overline{\rho u_i''\Phi} = \frac{\mu_t}{\sigma_{t\Phi}}\frac{\partial \tilde{\phi}}{\partial x_j} \qquad \text{E 6.19}$$

where $\sigma_{t\phi}$ is a turbulent Schmidt number for variable ϕ, typically taken to be 0.7–1.0. In the above set of equations the turbulent viscosity μ_t can be obtained from the relationship

$$\mu_t = \left(\frac{C_\mu\overline{\rho}k^2}{\varepsilon}\right) \qquad \text{E 6.20}$$

The set of constants used in the above turbulence model is available (Bray, 1996).

In case of a strongly radiating flame, an additional equation has to be solved for the enthalpy, neglecting body forces and viscous dissipation

$$\frac{\partial}{\partial x_j}\left(\bar{\rho}\tilde{u}_j\tilde{h}\right) = \frac{\partial}{\partial x_j}\frac{\mu}{\sigma}\left[\frac{\partial \tilde{h}}{\partial x_j} + \sum_{i-1}^{N}\left(\frac{1}{Le_i}-1\right)h_i\frac{\partial \tilde{Y}_i}{\partial x_j}\right] - \nabla \cdot q_r \qquad \text{E 6.21}$$

where the enthalpy of each species i may be expressed as

$$h_i(T) = H_i^0 + \int_0^T C_i \, dT \qquad \text{E 6.22}$$

The final major equation completing the formulation is the equation of state. We have for ideal gases

$$P = \frac{\rho R_0 T}{W} = \rho R_0 T \sum_{i-1}^{N} \frac{Y_i}{W_k} \qquad \text{E 6.23}$$

where R_0 is the universal gas constant. One of the significant simplifying assumptions in the treatment of turbulent reacting flows relates to the pressure being thermochemically constant. This assumption is applicable for low-speed flows but not for high-speed conditions. The constancy of the pressure in this manner does not imply that the fluid dynamics effect of the pressure may be neglected. Indeed, the mean pressure gradient and pressure fluctuations play a crucial role in determining the behavior of turbulent flows. In particular, coupling between mean pressure gradients and density variation arizing from heat release in turbulent reacting flows of gases result in countergradient diffusion (e.g., Bray, 1996).

Models for Turbulence-Combustion Interaction

In this section, a brief review of the state of the art in turbulent-combustion modeling, as described in the literature, has been presented. The specific topic of interest here was the modeling of turbulent gaseous non-premixed flames

Eddy Breakup Models (EBU)

Many calculation methods use an "eddy breakup" (EBU) model for the fast chemistry reaction assumption (Magnussen and Hjertager, 1977). These models are heuristic in origin but have been used with some success in many engineering applications and, in particular, gas turbine combustion chamber modeling. This success is attributed to the fact that the heat release in the gas turbine combustors is restricted to a narrow region in the primary zone where fast chemistry assumption might be applicable. The average formation rate may be expressed as

$$\overline{w}_{F,EBU} = -A_1 \overline{\rho} \frac{\varepsilon}{\kappa} \overline{Y}_L \quad \overline{Y}_L = \min\left(\overline{Y}_F, \frac{\overline{Y}_0}{r}\right) \qquad \text{E 6.24}$$

The modeling constant A_1 is seen to be a function of the form of the mixture fraction pdf and the value of stoichiometric mixture fraction; EBU models can be put on a sounder bases by modeling the dependence of A_1 on stoichiometric mixture fraction. This model appears to provide suitable solutions within limited range of variables, but it cannot be regarded as better than a rough guide and it should be recognized that the eddy breakup type of models, through intuitively correct, do not give rise to unique solutions.

Presumed pdf Shape

Turbulent reactive flows have been generally considered as an ensemble of random flowfields. Each of those fields does satisfy the classical thermochemical balance equations with a particular set of initial and boundary conditions. To specify these random fields, one needs to know the statistical information such as mean values, variances, covariance, and if possible, the pdf. The knowledge of the pdf is necessary to determine the scalar mean values. The conventional average of any scalar quantity (can be determined as

$$\overline{\phi} = \int_0^1 \phi \overline{P}(\phi) \, d\phi \qquad \text{E 6.25}$$

Applications to Coal Combustion

Again, there are innumerable applications. For example, the IFRF have obtained excellent results for many different coal flames in their burner test rigs. Thus, Visser and Weber (1990), for example, have modeled both unstaged and staged flames successfully. There are hundreds of papers describing the successful application of CFD to coal combustion. Examples are those described by Fiveland and Wessel (1991), Fiveland and Latham (1997), Jones et al. (1998), Stopford et al. (1998), Chen et al., 1992, and in recent reviews by Niksa (1996) and Abbas et al. (1996).

In combusting flows there are many more submodels and constants than in isothermal flows, and therefore greater uncertainties. For example, coals vary tremendously, and coals are often blended with the result that their properties, particularly for devolatilization and char oxidation, are often not very well known. Also, although scattering can be taken into account in the radiation models, very little is known about the scattering cross sections.

6.9
Computation of Emissions

CO_2 is formed in the combustion of any fuel-containing carbon. When the combustion is complete, all the CO formed as an intermediate product will have been converted to CO_2 with the release of additional heat. When this is not the case and CO is important, an equation for it can be added to all the others with an oxidation reaction scheme. Most other emissions of environmental concern are minor species and do not affect the main combustion process during the normal operation of burners and plant.

Emissions of SO_x, consisting mainly of SO_2, depend entirely on the sulfur content of the fuel and on the effective mixing of any additives (such as limestone, $CaCO_3$) that may be introduced in, for example, fluidized beds to convert the sulfur into other compounds such as calcium sulfate.

NO_x formation is more complicated and has already been discussed. However, it also is a minor species and so is usually calculated in a post-processor after the main combustion calculations have been completed.

Soot can be calculated in flames, although not accurately in many large industrial flames. Particulate matter such as ash is usually collected at, for example, the bottom of the boiler, and smaller particles are removed in the gas cleaning by for example filters or electrostatic precipitators. In coal flames there is a complicated, poorly understood interaction between particulate and NO_x formation. Other particulates, such as acid smuts, may be formed as a consequence of the combustion process and are thought to be formed in the cooling flue gases, as are other trace species, such as dioxins and furans, but these are extremely difficult to predict.

6.10
Modeling of Combustion Plant

Mathematical modeling of combustion is now a major commercial operation, and there is intense competition between the different vendors of commercial computer codes. In view of the wide range of combustion plant and its use worldwide, it is rather surprising that commercial heat balance codes have been developed almost exclusively for fired process heaters, and CFD, although of general applicability, has usually been focused on modeling burners and coal-fired power station boilers. Of course, many companies have their own in-house programs, but these are not usually available externally, and in fact, in spite of the claims made, it is nearly always impossible to find out what is in them and how they have been validated. In addition, there are some commercial codes for specific processes, in particular, thermal radiation and complex chemistry.

Stationary combustion plants are one of the major applications of CFD. Early applications included simple single burners integral with combustion chambers. Attention then focused on more complex burners with air staging. Currently, work has concentrated on much more complex combustion plant involving a number of burners, in the case of a power station combustor this may involve 48 burners. In large combustion plant, burner-burner interactions are present and these cause many problems. There is a lack of detailed plant data available to validate these complex systems. Most of the illustrations so far have been for single burners in test rigs.

Choice of the Combustion Model

Despite the availability of complex chemical kinetic schemes that may contain 1000 reactions or more, most CFD models involve extremely simplified chemistry, for example, NO_x formation being added in a post-processor. The formation of slag can also be handled in this way, but soot has to be an inherent part of the main program.

As far as the combustion process is concerned, the key issue is the nature of the flame and the type of mixing. If combustion is turbulent, a simple Magnussen model may be sufficient, but for accurate modeling of NO_x formation a flamelet model may have to be chosen. For the chemical models simplified steps may be considered, for example, a single-step model:

$$\text{Fuel} + O_2 \rightarrow CO_2 + H_2O$$

but here the flame temperatures are for an "undissociated" flame and tend to be too high. If a two-step model is used:

$$\text{Fuel} + O_2 \rightarrow CO + H_2O$$

$$CO + \tfrac{1}{2}O_2 \rightarrow CO_2$$

then CO as an intermediate is included and a closer approximation to the dissociated flame temperature is achieved — but rarely are both correct. Comprehensive kinetic models can be used in flamelet models for diffusion flames, but here a library of data has to be produced from routines based on "opposed" diffusion flames — such as the SANDIA OPPDIF. An alternative method is to use "reduced" chemical schemes, e.g., Peters and Weber (1993).

The original coal combustion models were based on an oil-droplet model. At the present time there is considerable interest in modeling coal combustion to improve low-NO_x pf coal burners [Jones et al. (1997), Niksa (1996), Williams et al. (1997), Tilley et al. (1998), Niksa et al. (1999)] and other NO_x-reduction techniques based on combustion modifications such as furnace air staging and furnace fuel staging (i.e., reburn). In most current CFD codes the process of modeling coal combustion is usually simplified to the following reactions set out in Chapter 5:

Step 1: $\text{Coal} \rightarrow \text{char} + \text{volatiles}$

Step 2: $\text{Volatiles (HC)} + O_2 \rightarrow CO + H_2O$

Step 3: $CO + \tfrac{1}{2}O_2 = CO_2$

Step 4: $\phi\, C\,(\text{char}) + O_2 \rightarrow 2(\phi - 1)CO + (2 - \phi)CO_2$

Step 1, devolatilization, is often simplified by using a first-order global rate with a single Arrhenius expression.

For bituminous coals, volatile yields are much greater at higher temperatures and depend strongly on heating rate. Consequently, a two-reaction devolatilization mechanism is required such as that developed and quantified by Kobayashi et al. (1976).

The rate parameters for k_1 are determined from the proximate analysis, while those for k_2 are determined from high-temperature pyrolysis. The product yield of chars and volatiles (and their nitrogen contents) and their rate of formation may be determined by devolatilization models such as FG-DVC (Solomon and Fletcher, 1994) or FLASHCHAIN (Niksa, 1996).

The char combustion mechanisms likewise are those set out in Chapter 5. Other models include the unreacted shrinking-core model, which can incorporate ash diffusional resistances to oxygen. Catalytic effects of the ash have received less attention, but these are mainly of significance at lower temperatures, yet could be effective at fluidized-bed-combustion temperatures. An alternative approach is the use of empirical correlations that relate char reactivity to experimentally measurable char or coal properties. This approach is often limited to a particular suite of coals/chars and so lacks accurate predictive applicability.

It is desirable to be able to model porous char formation and char structure, since these influence its combustion properties. Thus, areas for potential refinement within char burn-out models include coal structure and maceral effects on porosity, fragmentation, and reactivity of chars produced under different temperature-time histories. Such a model might incorporate a statistical distribution of char-particle reactivities such as that developed by Hurt and Davis (1994) and Hurt et al. (1996). In this, a statistical model (the CBK model) is used to describe the variations of reactivity of char particles remaining near 100% carbon burn-out. Equations for char combustion become increasingly inaccurate as combustion proceeds because of the hindering effects of ash and the reduced reactivity because of the graphitization of the carbon. A further aspect that can be considered is slag formation. Such models are very much in the research stage, but considerable developments will occur over the next few years.

In addition, an ideal model would treat the interrelationship of these factors and devolatilization with char-N, and hence, NO_x release during char combustion. The basic features of NO_x modeling are outlined below.

NO_x formation in coal flames is dominated by the fuel-N route, and the methods used are those set out earlier in Chapter 5.

The formation of soot and its burn-out are difficult to model accurately. One of the earliest models is the global model of Khan and Greeves (1974), adapted for oil burning by Abbas and Lockwood (1985). A more detailed model has been developed by Moss and co-workers (Moss et al., 1988) and by Fairweather et al. (1992). These models and their application to cone flames has been reviewed by Brown and Fletcher (1998).

The combustion of biomass and many waste materials can be modeled by the same techniques used for coal, and this is done for co-firing. The detailed mechanism and rates do differ and more information is required here.

6.11
Power Station and Other Boilers

Current interest is concerned with complex coal models of the type already outlined in multiburner combustion chambers. Details are also required, for design purposes, of heat transfer, slagging, and NO_x formation.

Very considerable advances have been made in the gridding of complex burners and of multiburner furnaces. An example of a complex power station boiler is given in Figure 6.14, the associated burner array is shown in Figure 6.15, and a burner in Figure 6.16. Figure 6.17 shows the temperature profile in a utility boiler, and this is typical of a whole host of computations that can be carried out in combustion chambers including opposed firing. Gas and particle velocities and their trajectories for groups of particle size ranges can be computed, and this is illustrated in Figure 6.18. A tremendous amount of effort has been put into computations of this type in the last 5 years, and with the advent of more powerful computers anything is possible — almost.

Nitrogen oxides arise from three routes, namely, thermal (reactions of N_2 with O at high temperatures), prompt (reaction of N_2 with fuel radicals), and fuel (release of fuel nitrogen during combustion), as previously described. In many CFD models these last two routes are represented by global reactions. The fuel-NO has a contribution from both the volatiles and the char nitrogen and it is assumed that all volatile nitrogen is released as HCN; an empirical relationship is used to estimate the fraction of char-N converted to NO_x. A more accurate prediction on NO_x release can be made if the full NO_x chemistry is used. Such an example of coal combustion in a single burner is given in Figure 6.19, where details are given of coal-particle trajectories, char burn-out rate, and predicted NO concentration in the coal furnace. In Figure 6.17 details are given of the effect of firing pattern and flow patterns within an opposed wall-fired furnace.

There have been many other examples of application of coal combustion models with complex burners and furnaces and the effects of combustion parameters, e.g., Fiveland and Wessel (1991), Lockwood and Romo-Millares (1992), and Abbas et al. (1996) have calculated the correct trend in the variation of NO with coal-particle size. Chan et al. (1983) have obtained good results for carbon burn-out and Stopford et al. (1994) have explained maldistribution in the exit plane temperature of a utility boiler. However, it is extremely difficult to predict very accurately the unburned carbon (carbon in

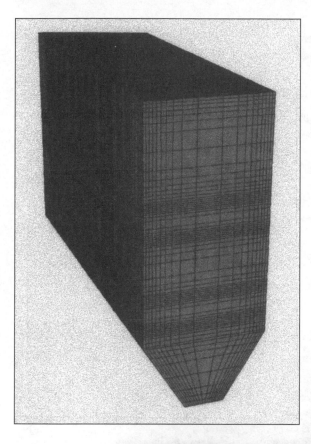

Figure 6.14 A multiburner power station combustor (Fluent, Inc.).

Figure 6.15 Furnace simplified burner model (CFX).

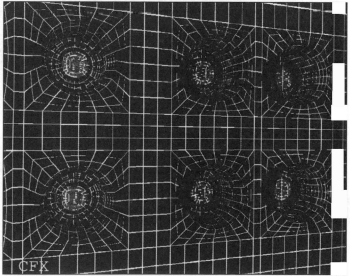

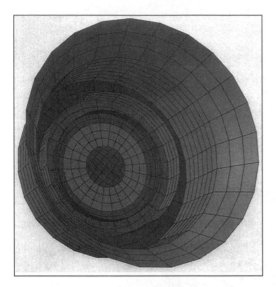

Figure 6.16   Low NO$_x$ burner grid
(Fluent Inc.).

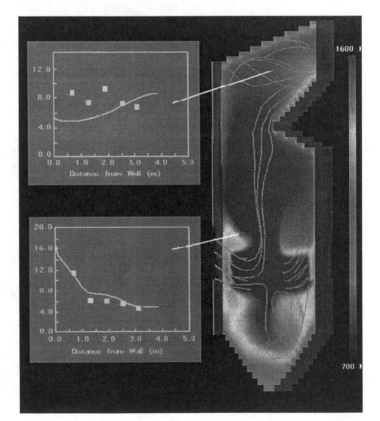

Figure 6.17
Computed tem-
perature profiles
in a utility boiler
(Smoot, 1999).

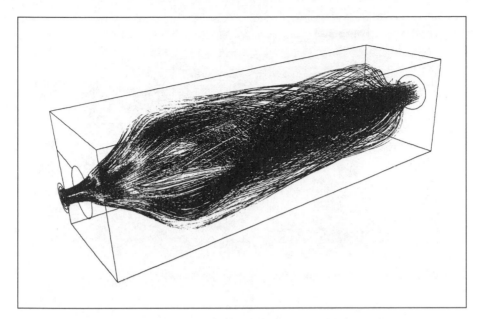

Figure 6.18 **Pulverized coal-particle tracks in a large-scale burner test facility. Improvements to the input allow the user to specify coal properties in an industry-standard format (Fluent Inc.).**

ash) for different types of coal, especially blends, while air ingress into combustion chambers frustrates many predictive methods.

6.12
Fluidized Beds and Stokers

These are extremely difficult to replicate. In the case of fluidized beds the present models cope can only with the noncombustion heat transfer case. While there are a number of theoretical models of fluidized beds, e.g., Kunii and Levenspiel (1991), Kulaekavan and Agarwal (1998), and Norman et al. (1997), they are capable of producing data only on the gases produced and are relatively poor at predicting the burning rates of bed material. However, considerable progress is being made, and Figure 6.19 shows the application involving the formation of gas bubbles in a fluidized-bed riser, calculated by a commercial code; such methods are also applicable in coal gasification.

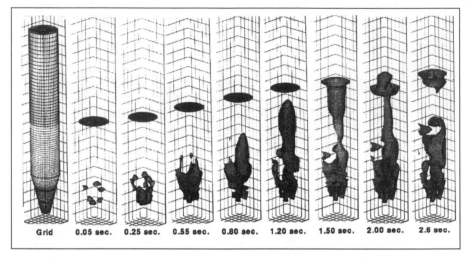

Figure 6.19   Computations using CFD methods in a fluidized bed (CFX).

Stokers (and fixed beds) present many complex problems with the definition of the "bed," and it is possible, at present, to model only an idealized model using a porous-media model. It is relatively easy to model gas-phase reactions above the bed, and this has been undertaken, for example, by the Swithenbank and Nasserzadeh Group for the combustion of municipal waste and for clinical waste (Nasserzadeh et al. 1993). This has been used to calculate temperatures and velocity streamlines in a domestic waste incinerator. This model is based on processes occurring in coal and involves the steps of devolatilization, char oxidation, and NO_x formation. The computation giving typical flow streamlines and temperatures and it represents the waste material as a large number of solid fuel particles located on the grate.

7

TWO-COMPONENT COAL COMBUSTION

7.1
Co-firing of Coal with Biomass or Waste

Materials such as biomass and waste (refuse) are solid fuels in their own right, but of low calorific value, of variable composition, and requiring special low-intensity combustion conditions. If mixed with coal, the average calorific value is increased, enabling it to be burned in conventional combustors and to have more average fuel properties. If coal is mixed with material such as a typical biomass, then the combustion process can be considered to be increasingly CO_2 neutral. If coal is mixed with refuse or other waste materials, its combustion provides a route to the disposal of unwanted materials with the advantageous production of energy; hence, this route is known as "energy from waste." Sometimes the distinction between waste and biofuels is blurred and some waste has a composition akin to a biofuel.

There have been extensive research programs over the last decade to increase the use of biomass, and one of the easiest ways of introducing this technology is by co-firing with coal in existing coal installations. A similar approach can also be taken with the disposal of waste, especially domestic refuse. Very extensive research studies have been undertaken, especially in Europe, that have been supported by the research programs Joule, APAS, and Thermie (e.g., Bemtgen et al.,

1995). This work has also been reviewed by Davidson (1997). Coal is derived from biomass by the process of coalification, and therefore biomass is essentially an extremely low-rank coal. The compositions of a range of diverse biomass materials are shown in Tables 7.1 and 7.2. The tables illustrate the high oxygen content and low carbon content of biomass compared with hard coals.

The combustion of biomass therefore also parallels that of coal in many respects. The first step involves devolatilization similar to that described for coal (Chen et al., 1998), namely:

$$\text{Biomass} \ \rightarrow \ \text{volatiles} \ + \ \text{char} \qquad\qquad \text{R 7.1}$$

where the volatiles have a composition controlled by the biomass composition, the heating rate, and the final temperature. A typical representative biomass would be cellulose, which is the most abundant organic substance in the world, 90% of which is in trees. Cellulose consists of glucose molecules ($C_6H_{12}O_6$) linked via oxygen bands, hence the high oxygen content in biomass fuels. Many biomass materials contain mainly CO and some hydrocarbons; consequently, the composition of the volatiles is dominated by the production of CO, some hydrocarbons, and a small amount of tar. The char itself has a reactivity very similar to coal char and does not contain as much ash, although the nature, their metallic content of Na, K, can mean that they have an enhanced catalytic reactivity especially at low temperatures.

Very extensive measurements have been made of the rate of the pyrolysis step R7.1 under low-temperature pyrolysis conditions; few comparisons have been made at high temperatures. However, some estimates have been made at high temperatures, and essentially it may be deduced that this step is not dissimilar to a typical bituminous coal such as Pittsburgh 8.

As far as mixtures of biomass and coal are concerned it can be concluded that

TABLE 7.1
CARBON, HYDROGEN AND OXYGEN COMPOSITIONS
FOR SELECTED BIOMASS COMBUSTION FUELS

Biomass	Carbon (% daf)	Hydrogen (% daf)	Oxygen (% daf)	Chlorine (µg/g)	Nitrogen (% daf)
Wood	45–50	5–6	40–45	0.04–0.06	0.3–0.5
Straw	40–45	5–6	40–45	0.34–0.36	0.5–0.7
MSW	30–35	1–2	20–25	0.15–0.20	1.0–1.5
Sewage sludge	40–45	5–6	20–25	0.10–0.15	3.5–4.0
Miscanthus	50–55	4–5	40–45	0.16–0.18	0.4–0.6

TABLE 7.2
ANALYSIS OF FUELS (DRY)

	Typical UK hard coal	Beech	Pine	Straw	Miscanthus	Bark	Wood
Volatiles	37.1	83.2	82.1	78.8	78.2	67.2	47.94
Ash	5.0	0.34	0.45	3.66	4.9	1.6	0.75
Fixed C	53.2	16.5	17.5	17.6	17	22.2	9.82
C % wt daf	85.1	48.7	53	47.4	50.7	47.1	47.62
H	5.9	5.7	4.8	4.5	4.4		6.11
O	5.7						44.34
N	2.12	0.13	0.11	0.5	0.3	0.43	0.65
S	0.84	< 0.05	< 0.05	0.1	0.2	0.26	
Cl	0.41	< 0.1	< 0.1	0.4	0.2	0.02	
HHV (MJ/kg)	33.5	18.5	19.3	17.09	18.0	18.7	9.55
Ash hemisphere °C	1080	1420	1110	1140	1170		
Ash analysis (% wt. ash)							
SiO_2	31.47	15.2	28.6	56.2	70.6		6.45
Al_2O_3	17.6	2.65	2.5	1.2	1.1		0.66
TiO_2	0.6	0.26	0.1	0.06	0.06		0.02
Fe_2O_3	23.2	3.8	6.5	1.2	1.0		2.05
CaO	12.5	37.3	35.8	6.5	7.5		73.05
MgO	0.6	8.5	5.2	3.0	2.5		0.54
SO_3	2.6	3.0	3.0	1.1	1.7		1.93
Na_2O_3	4.2	3.0	1.9	1.3	0.17		0.17
K_2O	1.5	8.6	9.2	28.7	12.8		4.5
P_2O_5	6.6	13.7	3.3	4.4	2.0		1.90
mg/kg ash							
Sb		< 20	< 20	< 20	< 20		
As		< 30	< 30	32	< 30		
Ba		1080	1610	90	70		
Pb		185	170	45	< 30		
Cd		1.2	1.0	3	0.5		
Cr		495	260	58	< 30		
Co		95	115	< 20	< 20		
Cu	1530	385	85	< 30			
Ni	605	< 30	45	< 30			
Hg	< 5	< 5	< 5	< 5			
Se	< 30	< 30	< 30	< 30			
V	84	25	28	38			
Zn	550	320	125	226			

the initial components devolatilize and their chars burn independently. The volatiles mix, however, and the calorific value of this mixture and resultant flame temperature is very much a function of the composition of the components; the combined flame temperature is fed back to the initial chemical steps and, consequently, there is a very significant synergistic interaction. The subsequent rates of burn-out of the chars is determined by their individual reactivity (and combined temperature), and generally biomass chars burn out more rapidly than coal chars.

The same approach can be taken with waste. The types of waste suitable for co-firing include biomass waste, municipal solid waste, and automotive tires. The composition is variable but has a calorific value between 10–20 MW/kg, depending on the extent of recycling to which it has been subjected (i.e., paper, plastic, cardboard removal). Its nonhomogeneous nature means that it is more suitable for fixed- or fluidized-bed combustion, or gasification, rather than for use as pulverized fuel. However, if the waste is taken from a controlled source, e.g., waste paper, with a constant calorific value, then the process of co-firing is simplified. In the other extreme, relatively small amounts of certain toxic wastes can be co-fired with coal in cement manufacture. The cement contains the trace elements in a generally satisfactory way — although great care has to be used to ensure that it meets the required environmental emission standards.

Combustion Techniques

Grate and fluidized-bed combustion   Traditionally, biomass such as wood has been burned on fixed grates and this can be undertaken easily; this is the case when co-fired with coal, although the ash throughput will be greater.

A similar situation applies to fluidized-bed combustion. Indeed, fluidized-bed combustion was designed with the combustion of low-grade, low-specification fuels in mind. However, this has not been an extensively followed-up practice other than for research combustors for both technical and economic reasons. The principal use of fluidized-bed combustion is in China, but this does not use mixed coal-biomass combustion to any great extent, at least at present.

Pulverized-fuel combustion   The combustion of pulverized biomass and mixtures with coal is different from the above cases involving fixed beds, because most of the existing power plants are designed as pulverized-fuel plants. Modification of these plants for multifuel use is obvious but the difficulties are largely unknown—handling, storage, milling processes can cause problems as well as corrosion and slagging effects. Co-combustion of solid biomass in coal-fired power plants seems to be a promizing technique for the future to contribute both to the reduction of greenhouse gases and to the solution of the waste disposal problem.

Extensive experiments in test rigs and plant have been made to investigate the behavior of blended fuels in pulverized-fuel combustion (Bemtgen et al., 1995). Apart from operational and technical considerations, the emissions of gaseous and particulate matter were examined. Straw, miscanthus, and sewage sludge were co-fired at different ratios with hard coal and brown coal. The biofuels could be prepared with different particle-size distributions, and the influence on emissions and burn-out was investigated. Intensive in-flame measurements had been made to explain the detailed combustion course of co-combustion. At the end of the chamber, emissions of NO_x, CO, CO_2, SO_2, C_mH_n, and the burn-out were measured in all the flames. Some selected flames could be measured regarding the N_2O-, PCDD-, PCDF-, PAH-, heavy metal, and HCl-concentrations, and the slagging behavior. For a complete understanding of the reaction course, mathematical simulation of co-fired flames could be done. It was shown that the combustion of biomass in a pulverized-fuel rig is technically feasible with moderate milling energy.

The burn-out and CO emissions are within the variation of pure coal flames. The NO_x concentrations may be very effectively reduced with different primary measures due to a high-volatile and a low-N-content of biofuels. The low-sulfur biomass (straw, miscanthus) distinctly reduces the SO_2 emissions, whereas sewage sludge with a high sulfur content increases them. Due to the high combustion temperature the N_2O emissions are lower than 20 mg/m^3.

As expected from biomass fuel analyses, the HCl emissions are increased by adding biofuel. Hence, dioxin emissions can also increase with higher fuel-Cl input, but only if mixing is poor, permitting the incomplete combustion of carbon. Straw showed an intensified slagging tendency, whereas miscanthus slagging was only a little higher than coal slag formation.

With the help of in-flame measurement and mathematical simulations it can be shown that the reaction of straw and miscanthus is slower overall compared to coal, whereas sewage sludge reacts in an initial step.

Gasification

Biomass and wastes can be gasified with coal to yield gas and indeed liquid fuels or chemical feedstocks. There is a considerable amount of knowledge developed over a number of years concerning gasification and this technology can be simply extended to incorporate biomass and wastes. This is described in a number of detailed reports especially the programs conducted under the Joule-Thermie program of the European Union (Bemtgen et al., 1995).

7.2
Co-firing with Natural Gas

Co-firing pulverized coal with natural gas offers a method of reducing SO_2 and NO_x emissions and improving a combustion system's performance, thus offering a potential to reduce CO_2 emissions [Doing and Morrison (1997), Pratapas and Holmes (1990), Mason et al. (1998)]. By replacing a fraction of coal with natural gas, SO_2 emissions can be reduced pro rata since natural gas contains virtually no sulfur; likewise, particulate emissions are reduced. If the gas is injected into a low-NO_x burner's primary zone, the subsequent reduction in oxygen concentration enhances the staging effect of the burner, improving NO_x reduction capability.

Combustion problems experienced with retrofitted low-NO_x burners, such as flame impingement, and low-load flame instabilities can also be alleviated by co-firing, since the added gaseous fuel readily ignites upon entry into a furnace, resulting in a high-temperature region in the near burner zone that stabilizes and shortens the coal flame by reducing ignition delay of coal particles (Bayliss et al., 1994).

Studies have shown that the co-firing of coal with gas is not as beneficial as reburn with natural gas in terms of NO_x reduction, but there are the other operational benefits in terms of efficiency, SO_x, and particulate reduction. Co-firing with blast furnace gas has been studied, but in that case there was an increase in emissions of unburned carbon in ash because the larger volume of fuel gas changed the coal-particle trajectories (Ma and Wu, 1992).

7.3
Combustion of Coal-Water Slurries

Development of Coal-Water Slurries

As a result of the oil price rises in the early 1970s, a move "back to coal from oil" resulted in many countries. Many coal conversion programs were instigated in oil-consuming countries to offset the dependence on highly priced petroleum imports, and one outcome was the development of coal-water slurries as an alternative liquid fuel (McHale, 1985, Thambimuthu, 1994). Interest since has been much reduced in most countries toward the end of the century because of the reduction in oil-fired plant and the increase in use of coal- or gas-fired systems.

However, research has continued at a reasonable level (e.g., see Fuel and Energy Abstracts) into the use of slurries partially because it is an easy fuel to transport by pipelines, etc., and partially because of its use as a gasification feedstock.

A number of clean coal-derived liquid fuels are available, the major ones being solvent-refined coals (SRC), coal-oil mixtures (COM), and coal-water slurries (CWS). Coal-water slurries are also referred to as coal-water mixtures (CWM) and coal-water fuel (CWF). Both coal-oil mixtures and coal-water slurries are suitable for heating and steam-raising applications, generally as an alternative to heavy fuel oil, although they can be used as engine fuels. Coarse (> 1 mm diameter) coal-water slurries can be used as feedstocks for gasification plants and as a feedstock for direct coal injection into blast furnaces. The attraction of the coal-water slurry is its complete independence of an oil supply; coal-oil mixtures are effectively a means of extending oil supplies but are less economically beneficial. Coal-water slurry is simply produced by mixing pulverized coal with water together with a small amount of a surfactant and a stabilizing agent. The coal-particle diameters used are generally about 100 μm or slightly less, but "micronized" coal-particle sizes are, as the name suggests, of the order of μm's. The resulting liquid looks like a black oil but feels like water to the touch; it smells like coal. There are two major reasons for investigating the suitability of coal-water slurries as a fuel. First, the fuel can be stored and burned in a similar way to heavy fuel oil in existing oil-fired appliances with only a few plant modifications, and second, coal-water mixtures can easily be transported in pipelines. However, there are some disadvantages to be overcome, such as increase of wear to pumps and atomizers due to the abrasive nature of the coal particles, blockage of mechanical components, flame instability, changes in heat transfer in the combustion chamber, as well as stability and flow problems during storage and pumping. The installation of a system to remove particulates such as ash and unburned carbon from the flue gases to meet environmental regulations may also be necessary; but it is possible to reduce NO_x by staged combustion, and to reduce SO_x by fuel beneficiation or in situ reaction with limestone present in the slurry.

The techniques that have been used for the utilization of heavy fuel oil combustion have been applied to the combustion of coal-water slurries. This is because after the pulverized coal has been mixed with sufficient water to achieve the required viscosity, there are certain similarities with heavy fuel oil with regard to both the rheology and to the combustion mechanism.

The concept of mixing pulverized coal with oil to form a fuel is not new, since the earliest patent on coal-oil mixtures is about 100 years old. Research into coal-oil mixtures was undertaken in the 1940s and afterward in the UK, Germany, Japan, and the US. The development of coal-water slurries is more recent: while some work on coal-water mixtures had been undertaken during the 1960s in Germany and Russia, the major developments were made in Sweden during the 1970s, and considerable progress was made in developing and marketing a coal-water fuel.

Slurry preparation   The key feature that has emerged from slurry-droplet combustion studies is that the small coal particles originally present in the slurry can agglomerate, especially in large droplets, resulting in chars larger in size than if the original coal had been burned directly as a pulverized fuel. The implication is that a longer combustion time is required than for pulverized fuels or oil.

Considerable attention, therefore, has been directed at both the slurry preparation technique and the atomization process in the hope that methods can be found in which the coal particles are thrown apart rather than agglomerating during combustion. This might be achieved by using small droplets or preheated slurries so that flash droplet vaporization occurs. Additives may be used that disruptively break up the droplets or that lower the viscosity of the fluid so that, when atomized, very small droplets are produced. The viscosity of the fuel and the quality of atomization that can be achieved are therefore crucial factors.

The relative proportions of coal and water that can form a suitable fuel are restricted by several factors. The coal must be present as a large fraction of the mixture, but in order to produce a practical flowing liquid fuel capable of being pumped and atomized, the presence of a certain amount of water is necessary. The highest coal loading that will still provide a slurry that satisfies operational requirements, particularly flame stability criteria, appears to be less than 80%. For this reason, most of the coal-water slurries that have been investigated so far contain between 60 and 75% by mass of pulverized coal. Not only coal loading but also the maximum size of the largest coal particles and the size distribution of the particles can affect both the rheology of the slurry and also the combustion efficiency, since the smaller coal particles have a more rapid burn-out. However, the economic benefits in combustion efficiency and ease of slurry handling gained by grinding the coal to a finer size are offset to a certain extent by the higher cost of pulverizing the coal. Therefore, a typical coal-water slurry contains about 70% mass of coal pulverized to minus 200 mesh (i.e., 74 μm diameter) or finer, with about 1% mass of an additive. These are various dispersants, surfactants, neutralizers, or stabilizers that, when present in small concentrations, have the ability to alter the rheological properties of the liquid mixture and thus allow a larger coal loading. They also prevent the coal and water separating out by sedimentation during storage and prevent bacterial growth.

The first stage of CWS preparation is usually dry grinding of the coal with subsequent mixing with water. In wet grinding, the water and the dispersant can be added before grinding begins, and beneficiation of the coal by froth flotation methods can be included in the grinding stage. If a flocculant is added to the water and ground coal mixture, the coal particles can be made to settle while the suspended pyrite and mineral matter particles can be decanted. This process may be repeated if necessary, and low ash levels of ~ 2% and the removal of up to 90% of the sulfur may be achieved with some coals.

Coal can be cleaned to remove sulfur and ash by washing with water and by density separation and flocculation processes; i.e., it is potentially a clean coal technology. By incorporating the beneficiation process in with the wet grinding stage of slurry fuel preparation, the energy-absorbing, and thus costly, dewatering stage necessary, for dry coal cleaning, is no longer necessary and a reduction in the environmentally damaging sulfur and the boiler fouling ash is obtained.

Chemical cleaning can also be incorporated to provide an ultraclean coal-water slurry suitable for burning in boilers without the need to add flue gas pollutant removal systems. Both one- and two-step grinding techniques in ball mills have been used to produce the required particle-size distribution. The two-step process results in a bimodal particle-size distribution. Clearly, a high proportion of fine-sized particles will maximize the coal loading without raising the viscosity to an unacceptably high level, but the effect is offset by the increased cost of fine grinding the coal. The available surface area contributes to the rate of char combustion, and slurry droplets containing many fine coal particles may agglomerate to a more closely filled and denser structure than those that have a distribution of particle sizes. Combustion studies suggest that a CWS with coal particles with mass median diameter about 20–30 mm with about 70 or 80% less than 200 mesh (~ 50 μm) yields a higher combustion efficiency than do both finer or coarser slurries.

Slurries have been produced with various rheological properties depending on the nature of the additives used. The first requirement is that the coal is well dispersed throughout the water. Coal surfaces have both hydrophobic and hydrophilic sites, the relative proportions varying from coal to coal. In water alone, parts of the surface of the coal particles will be wetted only slowly and a hydrophobic agglomeration of coal particles can occur. A suitable surfactant decreases the surface tension of the fluid or modifies the electrostatic charges on the surface so that the particles are rapidly wetted and the coal will become well dispersed in the liquid. However, even in the presence of a suitable additive, the maximum coal loading that provides a low-viscosity slurry is about 60% with lignites and subbituminous coals and about 75% for bituminous coals.

The other rheological property required is that the fuel should have a low viscosity, so that it produces an easily atomized fuel. Coal-water slurries are non-Newtonian fluids (that is, the viscosity is not constant but is a function of the applied stress), and therefore, a suitable additive should lower the viscosity at high shearing rates and improve the thermal stability. Figure 7.1 shows the vari-

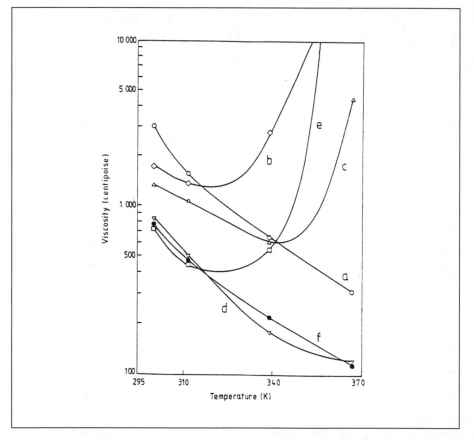

Figure 7.1   Variation of viscosity at low shear rates with temperature for six different CWS fuels (after Daley et al., 1985).

ation of viscosity with temperature at low shear rates for six different coal-water slurries from different suppliers. Some slurries show a reduction of viscosity with increasing temperature, while with others the reverse is true; this is thought to be due to the change in behavior at high temperatures of the additive used. Various types of additives such as ionic salts, gums, clays, and starches have been investigated. Ionic additives can have an adverse effect on combustion characteristics and corrosion, and nonionic substances such as long-chain fatty alcohol derivatives, especially polyethoxylates, are particularly suitable additives for coal-water slurries.

The presence of about 1% of an additive has been found to not greatly affect combustion performance compared with combustion of a similar slurry without the additive; carbon conversion efficiency is slightly lower with an additive present, and atomization characteristics and the emission in the flue gases of particulates and SO_2, CO, and NO_x are similar.

Atomization of Coal-Water Slurries

The atomization stage is the most critical part of the process of coal-water slurry combustion, since both ignition delay and combustion efficiency are dependent upon the size of the droplets in the spray. Smaller droplets result in smaller agglomerates, and also present a larger surface area for a given quantity of fuel than do larger droplets. This increase in surface area increases the rate of water vaporization, thereby reducing the ignition delay time and facilitating more rapid char burn-out. It is probable that in order to achieve the desired carbon conversion efficiency it is necessary only to limit the number of droplets that have a diameter greater than 300 μm. The fineness of the droplet-size distribution depends on atomizer design and the properties of the fuel — its viscosity, temperature, and mass flow rate of the atomizing medium. The smaller the outlet orifice the finer the spray produced (although this is limited by the larger coal-particle sizes), but also the greater the shear rate on the fuel necessary to achieve a particular firing rate; however, some CWS fuels have a tendency to dilatancy, or an increase in viscosity at high shear rates. It has been generally shown that fuel and atomizing medium preheat, and the dilution of the fuel with water, can improve atomization quality. However, atomization of slurries is a much more random and irregular process than the atomization of liquids. Sheet breakup results in large segments of liquid rather than well-ordered ligaments.

Atomizers designed for use with heavy fuel oils are not automatically suitable for use with CWS fuels since, in regions where high velocities and change in flow direction exist, there is a tendency for the surfaces to experience erosion after only moderate use due to the abrasiveness of the coal. There may also be clogging of any narrow passages by coal particles. Erosion can be reduced by replacing parts prone to erosion with harder materials such as ceramics. Several different types of atomizer suitable for use with coal-water slurries have been designed and tested, and four are shown in Figure 7.2. Atomizer (a) is a single-hole design in which slurry and swirling air are mixed internally in a chamber lined with alumina to resist wear. An internal mix "Y"-jet design with twin air streams is shown in (b); potential wear problems are reduced by using tungsten carbide in crucial areas. Atomizer (c) is an external mix atomizer designed

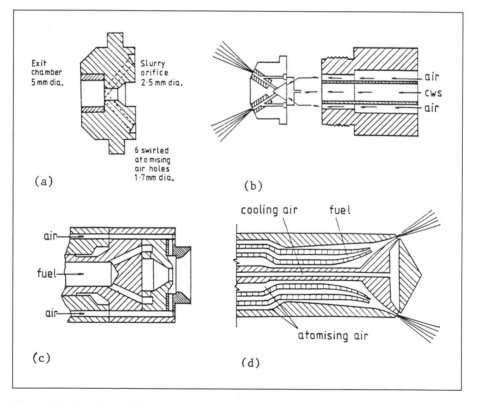

Figure 7.2   Atomizer designs suitable for use with coal-water slurries. (a) Single-hole, internal mix; (b) Y-jet internal mix; (c) external mix; (d) Lezzon nozzle (Murdoch and Williams, 1986).

specifically for use with coal-oil mixtures. The liquid leaves the swirl chamber orifice in the form of a thin hollow cone that is broken into droplets on meeting air flowing through tangential slots. The Lezzon design in (d) uses inner and outer air flows to shear a liquid-fuel sheet. The initial contact of these flows occurs inside the nozzle, and the fuel leaves the nozzle in an annular gap near the circumference of the tip. Spinning cup atomizers, of the type used for conventional fuel-oil mixtures, have also been used successfully.

Combustion Mechanism of Coal-Water Slurries

As the coal is heated in a flame on a burner, the water evaporates, which exposes the coal particles to the radiation from the flame gases. They undergo devolatilization with the emission of gases and tars, but the particles agglomerate via a bridging mechanism involving the coal tars produced. As the coal devolatilizes, it forms a carbonaceous skeletal structure (char) that subsequently undergoes surface combustion with the surrounding gases, leaving the mineral matter as an ash [Pourkashanian and Williams (1983), Walsh et al. (1984), Murdoch et al. (1984), and Takeno et al. (1996)].

The combustion mechanisms of oil and coal-water slurry are basically similar, and in the following the exact mechanism of coal-water slurry combustion is discussed further. Experimental studies in which a single droplet of a coal-water slurry undergoing combustion is suspended by a supporting fiber or thermocouple so that it can be observed and monitored continuously over the droplet lifetime have revealed details of the mechanism of slurry combustion. Therefore, the single-droplet studies have been used to provide much useful information about the behavior of various types of coal-water slurry as it has in the case of coal combustion (however, single coal-particle studies can yield extremely useful information).

Typically, a small droplet of slurry with a diameter of 1 mm or smaller is suspended on a fine silica fiber (or a thermocouple) and heated, usually by a furnace or laser. In order to monitor temperature changes, the droplet is suspended on the bead of a fine thermocouple and placed in a furnace maintained at a known temperature. The emission of light can be used to indicate the instant at which ignition occurs, and the behavior of the droplet can also be recorded by a movie or video camera. The rate at which the droplet mass decreases with time can be obtained by heating in a furnace at known temperature a slurry droplet suspended on a quartz fiber attached to the arm of a microbalance. Another technique is to allow a coal-water slurry droplet to fall through a heated vertical furnace and to observe the progress of combustion through viewing ports as the droplet falls. Figure 7.3 shows a typical temperature versus time curve, and Figure 7.4, the relative diameter and relative mass versus time curves for a single suspended CWS droplet. The slurry droplet follows the typical behavior of a heavy fuel-oil droplet and exhibits an ignition delay period t_I, a visible envelope flame period t_F, and finally, a char combustion period t_c, characterized by glowing char and by the absence of a visible flame, until eventually the char disinte-

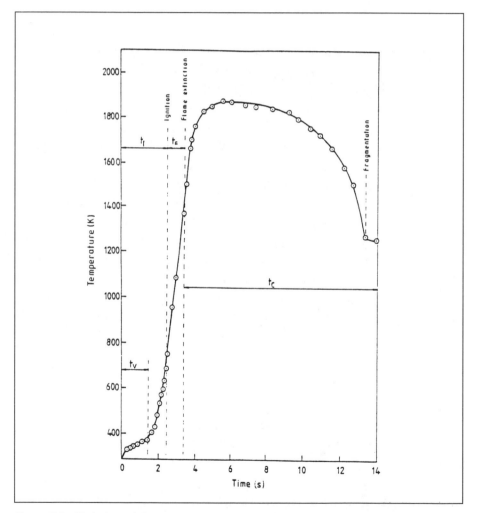

Figure 7.3   Variation of the temperature at the center of a bituminous coal-water slurry droplet with time (Pourkashanian and Williams, 1983).

grates into small fragments. Figure 7.5 compares and contrasts the differences between a fuel oil, Orimulsion, which is a commercial fuel consisting of 80% bitumen particles in water emulsion, and a coal-water slurry. During the ignition delay, the period t_t is associated with the droplet heating to the boiling point of water and the evaporation of water from the surface of the droplet. When most of the water has been evaporated, the coal particles heat up, leading to coal-particle devolatilization. If the droplet is small (say, ~50 mm or less) there may be only a small difference between the temperature at the center and at the surface

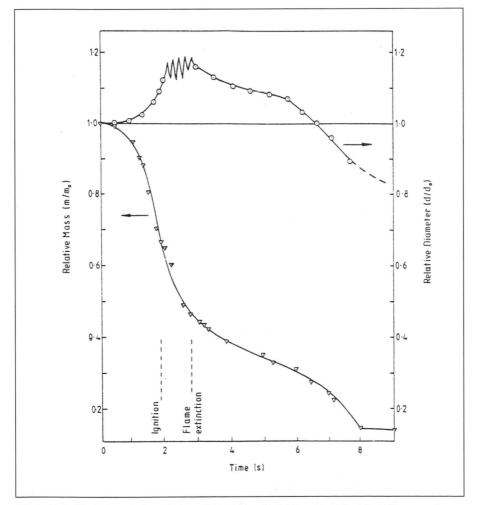

Figure 7.4   Variation of the relative diameter and relative mass of a bituminous coal-water slurry with time (Pourkashanian and Williams, 1983).

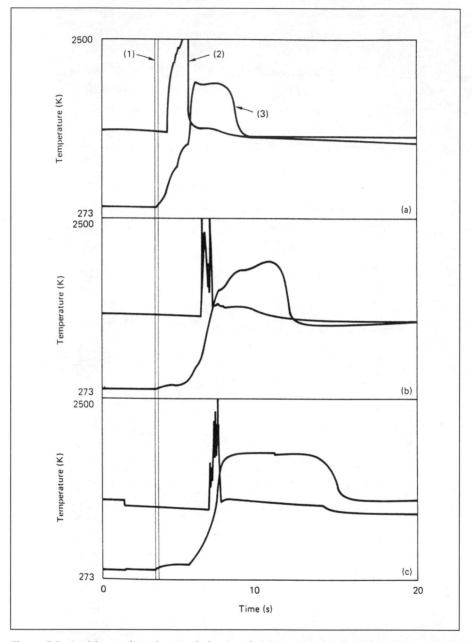

Figure 7.5   Ignition and combustion behavior of 1.1-mm droplets introduced into a fur-nace at 1115K of (a) medium fuel oil, (b) Orimulsion, and (c) coal-water slurry. The in-stant injection is indicated by (1); the light emission indicating the onset of ignition and the duration of visible flame is shown by (2); the droplet center temperature is indicated by (3), which also indicates the char burn-out time (Pourkashanian and Williams, 1987).

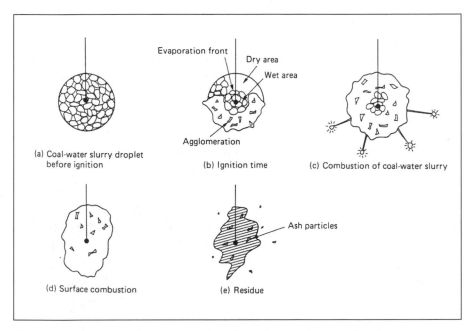

Figure 7.6   Diagrammatic representation of a coal-water slurry combustion.

of the droplet, and the water at the center will vaporize before the surface coal particles start to devolatilize. However, if the slurry droplet is quite large or the heating rate very high, the situation may be similar to that represented in Figure 7.6, which is a simplified diagram of the various stages of slurry-droplet combustion. Some water still remains in the center of the droplet, but the coal particles on the surface, being at a higher temperature than the center, will have started to devolatilize and swell, causing agglomeration in the surface region. The initial swelling is usually in the form of small bubbles produced as the volatile gases force their way out of the coal particle, but, in highly swelling coals, parts of the surface expand considerably while the particle surface is fluid. Meanwhile, the water at the center vaporizes and the pressure builds up until eventually the water vapor forces a way through any weak spots in the agglomerating surface, forming blowholes or craters as well as internal voids. Volatile products from the devolatilization stage of the coal particles situated at the center of the droplet may also form blowholes and voids in a similar manner when trapped by the swelling coal particles at the surface; a consequent increase in droplet size while the droplet mass is decreasing is clear in Figure 7.4. Some time during the period of evolution of volatile gases, ignition of the surrounding envelope occurs and a visible flame is formed.

When swelling ceases, which coincides with the extinction of the visible flame, the residual particles become more rigid and undergo a slight contraction as they form a char; this char then oxidizes, with its surface glowing red. The chars from highly swelling coals, being more porous, tend to burn more rapidly, due to the increase in surface area, than the chars from low-swelling coals which have a denser, more compact structure. Fragmentation also occurs during the later stages of char burn-out. Finally, only the ash from the original coal remains.

In large-scale spray flame situations, a range of droplet sizes is present, and the size and density of a droplet may affect its velocity and flow pattern within the combustion chamber. Clearly, the course of combustion may vary from one droplet to another depending on the size and position in the chamber, but the mechanism will essentially be similar to that observed for a single droplet.

Similar mechanisms apply to combustion in a fluidized-bed-combustor and indeed to the initial stages of the gasification of coal-water slurries in a gasifier.

Theoretical Modeling of Coal-Water Slurry Combustion

If a single coal-water slurry droplet is considered entering a combustion chamber, it heats up and vaporization of the slurry water takes place; this water contains a small amount of additive and possibly a few soluble components leached from the coal itself. The coal particles dry and agglomerate due to surface tension effects and due to the stickiness of tar liquids that are formed as the coal heats up to its plastic stage (~600–850K). Noncaking coals have also been found to agglomerate to a certain extent. The effect of agglomerization as far as devolatilization of char combustion is concerned is to decrease the surface area available. Photographic studies on coal pyrolysis in an inert atmosphere have shown that vapors and tar liquids can jet out of the walls of the coal particle and considerably increase the surface area by forming holes and craters; these jets can also cause the droplet to rotate.

Combustion of the char residue involves a number of reactions that may occur on the surface of the char. The reactions of char with water vapor, and also with carbon dioxide, are slow compared to that with oxygen, but at the temperatures reached in the vicinity of the agglomerated char, the presence of water vapor will serve as a catalyst to the oxidation of carbon monoxide in the gaseous phase.

The most important gaseous-phase reactions are those previously outlined for volatiles combustion. In the early stages of char combustion, depending on the initial droplet size, and while the agglomerated mass is large, the reaction rate will be controlled by diffusion. As the mass burns away and the temperature increases, the reactions will proceed more rapidly and will become chemically controlled.

Figure 7.7 gives the results of computation showing the effect of droplet diameter on the time taken for the droplet water to vaporize, the volatile flame to ignite and extinguish, and for char burn-out as calculated from a one-dimensional model that assumes there is plug flow along the combustion chamber (Walsh et al., 1984). Similar calculations obtained for various coal loadings of

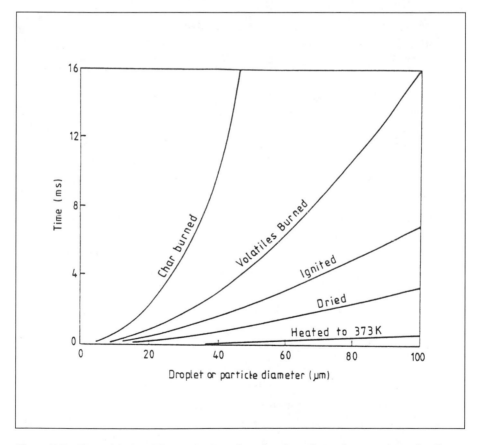

Figure 7.7 Times for the different stages of combustion of a coal-water-slurry droplet as a function of droplet size (after Walsh et al., 1984).

the slurry show that additional water acts as a heat sink and delays ignition, and that particle burn-out decreases slightly.

Coal-Water-Slurry Spray Flames

The main stages occurring in a combustion chamber are similar to those for oil. The fuel is transported from the storage tank, which may have to be agitated by slurries, by means of a pumping system, and after mixing with air passes through an atomizer. The fuel leaves the atomizer in the form of small droplets that are injected into the combustion chamber where they are burned. The mixing of fuel and oxidant depends on the spatial distribution and momentum of the gas-phase flow, as well as on the combustion chamber geometry. The fuel vaporizes and burns, followed by mixing and recirculation of the reacting hot gases. In the case where the liquid fuel is a coal-water slurry, the vaporization step is replaced by vaporization of the water, followed by the devolatilization and agglomeration of the coal particles, and the droplet burning replaced by the char combustion stage.

Fluidized-Bed Combustion of Coal-Water Slurries

The combustion of coal-water slurries in fluidized-bed combustors appears to be a very simple technique for a number of reasons. When burning dry pulverized coals in a fluidized bed there can be problems with feeding the crushed coal into the bed, particularly with pressurized fluidized beds. By comparison, the injection of the liquid slurry can be easily controlled and varied with changes in the firing rate and, as agglomeration of the coal particles occurs as the slurry liquid evaporates, there are less problems with elutriation. The extent of agglomeration depends on the type of coal used in slurry preparation, and there is a tendency for some of the agglomerated particles to sink to the bottom of the bed unless a bed material denser than coal is used. Slurries with coal-particle-size distributions that are coarser than that necessary for slurries used in other types of combustors have been burned successfully in fluidized beds. Combustion efficiencies of ~99% have been obtained with particle-size distributions in which the largest diameter is ~3 mm and with coals with a high ash content. Sulfur retention and NO_x emission levels are as good as when dry pulverized coal is burned. Since beneficiation and fine grinding are unnecessary in the preparation of CWS liquids for fluidized-bed combustors, and with the improvement in fuel feeding over dry coal, coal-water mixtures are ideal for this method of combustion.

Coal-Oil Mixtures (COM)

This technology by which mixtures of fine coal particles are dispersed in oil was developed in 1879, and there has been extensive research since then. Such mixtures have various terms including colloidal fuel, but more recently the term coal-oil mixtures has been used. While the fuel has been available for this long period of time, it has not been widely adopted for two reasons: first, it is not as easy to handle as either coal or oil alone, and second, the grinding process and stabilization process make it more expensive, although, from an environmental viewpoint, the coal does become beneficiated during processing. Although there is now a very considerable literature on COM, to all intents and purposes it has been replaced by coal-water slurries, which are much more economical although there are still some problems relating to handling, atomization, and combustion. Consequently, COM is dealt with here only in brief outline.

Production of COM

A suspension of powdered coal in oil is usually unstable and coal particles will deposit unless the mixture is "stabilized." Coal-oil mixtures thus involve a complex formulation consisting of about 50 wt% coal, 45 wt% oil, and a cocktail of stabilizing agents, particularly an electrolyte to provide a double layer of ionic charge around each coal particle to assist in stabilization. Choice of oil and fineness of the coal particles are also important parameters. The COM produced is thixotropic in a similar manner to coal-water slurries.

Combustion of COM

Very many combustion trials have been undertaken. Problems can arise in the handling system from (1) erosion because of the presence of the coal particles that can seriously effect pumps and atomizers, again increasing the cost, (2) the thixotropic and generally more viscous nature of the mixture, and (3) plugging of filters and settling.

Further problems arise during combustion. Although the droplets produced during atomization are comparable to that produced by conventional oil atomizers, and although the coal particles themselves are quite fine, it has been shown that during the combustion process the combustion time is longer than either for the oil droplets or the coal particles alone. The reason for this (Braide et al., 1979) is that as droplets become heated on entering the furnace, the oil component evaporates and the coal particles agglomerate. The evaporative stage involves some disruptive ejection of material, because as the surface temperature of the COM droplet increases, the coal particles themselves pyrolyze. However, the pyrolysing coal particles become coated with a viscous tar that links the coal particles by tar bridges. The agglomerated particle thus has sizes considerably larger than originally, with a consequential increase in burn-out time. Consequently, combustion equipment has to be derated. An attempt to overcome this has involved emulsifying water with coal-oil mixtures (40 wt% coal, 40 wt% oil, 10 wt% water) with the intention of causing disruptive combustion.

7.4
Formation of Briquettes and Smokeless Fuels and Their Use

Finely divided coal resulting from processing or coal-cleaning operations can be agglomerated into larger particles that are called fuel briquettes and that are now largely used only for domestic applications. The production of fuel briquettes has long been practiced, which can range from the simple binding of coal particles with a binder to sophisticated smokeless fuel briquettes (Pitt, 1979). Smokeless fuels can be produced from coke, but this type of fuel is not discussed here — the formation of coke and its properties is a major subject in its own right and is dealt with in a number of textbooks (e.g., Marsh, 1989, Patrick, 1995).

Briquettes can be formed from bituminous or subbituminous coals, mainly the latter (Speight, 1994, Gronhoud et al., 1982). There are two ways of producing briquettes. Generally, the first group involves a bitumen binder that is produced by the carbonization of a coal, although a bitumen produced from oil can be used. The second group involves briquetting without a binder. Generally, the following properties are required:

1. *Handling properties.* High density, therefore lower volumes to handle, resistant to abrasion, clean to handle, and uniform size with no fines
2. *Combustion properties.* Smokeless, reactive (ease of ignition — open fires), long-lasting (used in overnight-closed stoves), high efficiency (low C in ash), granular residual ash, and uniformity of performance

The various commercial processes involve:

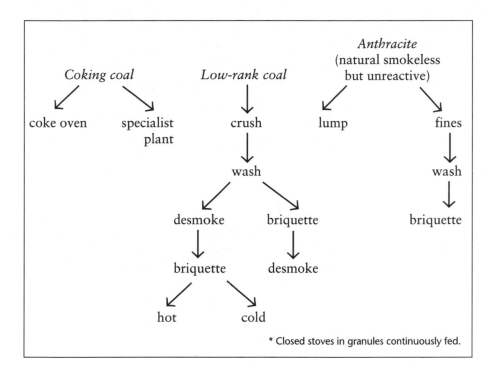

1. *Coalite process.* This has a higher heating rate → more fluidity. The coke-like material is removed just after it solidifies (lower *T*); some tars are also produced to give a valuable by-product.

2. *Desmoke then briquette (housefire process).* This involves a fluidized bed at softening temperature, minimum air present. The still-plastic material is extruded to give briquettes.

3. *Phurnacite process.* A briquette is first formed. The coal is mixed with a binder (pitch) having plastic properties and coal fines. It is then desmoked, carbonized in a tilted oven at 850°C, and decharged under gravity.

4. *Anthracite briquetting—Ancit process.* Here, powdered anthracite heated (600°C) coal is mixed with a binder coal (small amount). It is heated at about 300°C and the plastic composite is pressed into briquettes. It is then heated to remove sufficient volatile matter to be smokeless.

5. *Anthracite — mild heat treatment process.* Here, the binder is specially treated with a molasses + H_3PO_4 binder to stick anthracite together. It is then cured at 250°C and roll-pressed to give a hard briquette.

The combustion process involved with a briquette is fairly simple. The briquettes can be burned in fixed, traveling, or indeed, fluidized beds with ease. Essentially, the first stage of combustion is in part the reverse of the manufacture. The briquette heats up and the binder softens releasing the agglomerated particles. These then burn in close proximity, in a traveling grate they become dispersed, and in a fluidized bed they become fluidized. Briquettes can also be made from both coal and biomass and are used in some countries to burn fire coal and straw (or wood) mixtures.

8

COAL GASIFICATION PROCESSES

8.1 Coal Gasification

Coal gasification is a well-established technology in which coal is gasified with oxygen in an exothermic reaction to produce a combustible gas consisting of carbon monoxide, hydrogen, and some methane and carbon dioxide. This combustible mixture can be used as a fuel, and usually this is with a gas turbine to produce electricity, or it can be used to make a synthetic natural gas or chemical feedstock. Sometimes pure oxygen is used, in a process termed "oxygen-gasification," or air may be used in "air gasification" and then the product gas also contains nitrogen. Steam can be added to the oxidant stream to increase the amount of hydrogen in the gas produced; alternatively, water can be added as a liquid when the coal is injected in the form of coal-water slurry. Coal gasification has a wide range of applications that are set out in Figure 8.1, and a long history is associated with their development. Coal may be used as a feedstock, but the technique is applicable to any hydrocarbon, including natural gas and heavy refinery residues for example.

Coal gasification can produce a gas used for synthesis (syngas), or as a source of hydrogen for the manufacture of ammonia or hydrogenation applications in refineries, and many of the technologies have been developed by petroleum companies with these applications in mind. However, one of the main current interests

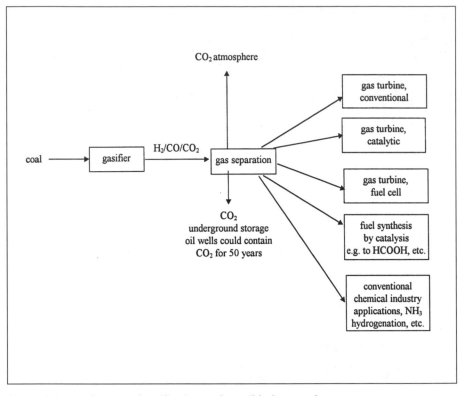

Figure 8.1 Application of gasification and possible future role.

is in the application of the gasified products in electricity generation. The conversion of coal to electricity, via such an intermediate gaseous product stage, can be achieved by employing the integrated gasification combined cycle (IGCC) technology. Here, a gasifier produces a fuel gas that is cleaned and then burned with compressed air in the combustor of a gas turbine to produce hot gases at high pressure. The gases are expanded through the gas turbine to drive the air compressor and a generator that produces electrical power. The hot exhaust gases from the gas turbine are subsequently used to raise steam in a boiler, which is then expanded through a steam turbine driving a generator to produce more electrical power. A typical arrangement of an IGCC is shown in Figure 8.2. Gasification is recognized as a clean and efficient alternative to coal combustion for power generation, and this could become an increasingly important aspect of coal utilization (Takematsu and Maude, 1991, Scott and Carpenter, 1996).

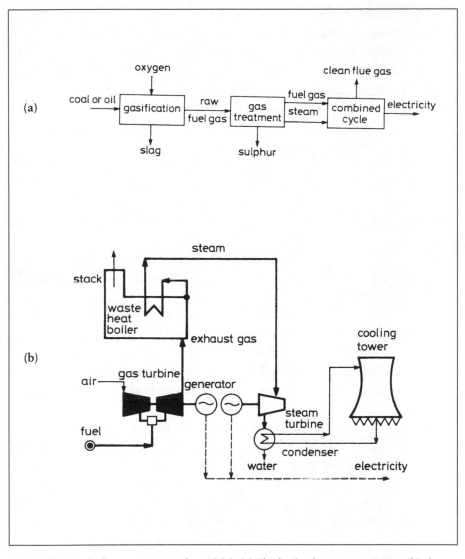

Figure 8.2 Typical arrangement of an IGGC. (a) The basic plant components; (b) the arrangements of the combined cycle system.

However, the state-of-the-art steam supercritical power plant with atmospheric pulverized coal firing is the yardstick for gasification plant. Current plant with a high level of pollutant emission reduction can achieve a net efficiency of about 43–45%, but supercritical power plant with state-of-the-art flue gas cleanup can achieve efficiencies of about 50%. Gasification plant has become a competitor only because of the development of large (250 MW), highly efficient gas turbines integrated with efficient thermodynamic cycles, so that in the long term efficiencies greater than 50% can be achieved with low emissions. Often, all the fuel that is gasified is burned in the gas turbine and the heat used to raise steam for the steam turbine cycles. Clearly, there is more scope for development, the use of bottoming and topping cycles, fuel cells, and the use of new high-temperature materials for gas turbines.

A further technological factor is that the gas turbines used are sufficiently rugged to accept a gas stream containing reasonably acceptable levels of corrosive gases and particulates. These challenges have been largely overcome, and consequently, there are 22 IGCC plants in or nearly in operation using a variety of feedstocks.

Whether air or oxygen is used as the oxidant determines the calorific value of the gasified gas. When air is used, the nitrogen in the air results in dilution of the product gas. Thus, the gasified gas will have a low calorific value (3–6 MJ/m^3 LHV) compared with coal gas manufactured using oxygen (10–12 MJ/m^3, LHV). This is much lower than natural gas, which has a calorific value of around 40 MJ/m^3.

On a worldwide basis most technological effort has been directed to the development of oxygen-blown gasifiers that produce a medium-CV gas and a high hydrogen content in the gas. The use of oxygen reduces the cost and size of the gasifier, auxiliary gas cooling, cleanup, and handling systems. The operation and carbon conversion of the gasifier also improves with use of oxygen. But, when compared to integrated gasification combined cycle systems, oxygen-blown gasifiers have one major disadvantage: they require an oxygen plant. An oxygen plant consumes about 5% of the gross power generated, which is the main reason why total plant investment for an oxygen-blown plant is somewhat higher than for an air-blown plant. However, in a fully integrated plant, the nitrogen (at the same pressure as the oxygen) is fed into the gas turbine with thermodynamic advantage. Further, this dilutes the fuel gas, resulting in a low calorific value gas but also lower NO_x formation in the gas turbine.

Tables 8.1 and 8.2 show the different types of gasifier that can be used. Apart from the differences in oxidant, there are differences in bed type that in turn influence the type of bed used (cf. Table 8.1), which in turn determines the size of coal used (cf. Table 8.2).

TABLE 8.1
THE DIFFERENT GASIFIER PROCESSES

Bed type	Coal feeding	Ash	Gasifier process
Moving	Dry	Dry	Lurgi dry ash
	Dry	Slag	BGL
Fluidized	Dry	Nonagglomerating	HTW (high-temperature Winkler)
		Agglomerating	U-gas (utility gas)
			KRW (Kellogg-Rust Westinghouse)
Entrained	Dry	Slag	Shell
			PRENFLO (pressurized entrained flow)
			VEW (Vereinigte Electrizitäts Werke Westfalen)
			GSP (Gaskombinat Schwarze Pumpe)
			Koppers-Totzek
	Slurry	Slag	Texaco
			Dow (Destec)

TABLE 8.2
IMPORTANT CHARACTERISTICS OF THE THREE TYPES OF GASIFIERS

	Moving-in bed		Fluid-bed	Entrained-flow
Ash conditions	Dry ash	Slagging	Dry ash	Slagging
Feed coal characteristics Size	5–80 mm	5–80 mm	5 mm	pf
Operating characteristics Operation temperature	673–973K	673–973K	1173–1373K	1473K
Key distinguishing characteristics	Hydrocarbon liquids in the raw gas		Large char recycle	Large amount of sensible heat energy in the hot raw gas

8.2
Basic Gasification Reactions

The coal gasification reactions occur when coal is heated with oxygen and usually some steam in a gasification reaction chamber. A typical bituminous coal has 77–90% carbon and contains 10–30% volatile matter, and it is convenient to discuss coal gasification mainly in terms of the reaction of carbon with a suitable gas. The main reactions which take place in a gasification system are

$$\text{Coal} \rightarrow \text{char} + \text{volatiles} \qquad \text{Endothermic} \qquad \text{R8.1}$$

$$C + \tfrac{1}{2}O_2 \rightarrow CO \qquad \Delta H^\circ{}_{298} = -123 \text{ kJ/mol} \qquad \text{R8.2}$$

$$C + O_2 \rightarrow CO_2 \qquad \Delta H^\circ{}_{298} = -406 \text{ kJ/mol} \qquad \text{R8.3}$$

$$\text{Volatiles} + O_2 \rightarrow CO \text{ (and } H_2O) \qquad \text{Exothermic} \qquad \text{R8.4}$$

$$CO + \tfrac{1}{2}O_2 \rightarrow CO2 \qquad \Delta H^\circ{}_{298} = -283 \text{ kJ/mol} \qquad \text{R8.5}$$

$$C + H_2O \rightarrow CO + H_2 \qquad \Delta H^\circ{}_{298} = -118.9 \text{ kJ/mol} \qquad \text{R8.6}$$

$$C + CO_2 \rightarrow 2CO \qquad \Delta H^\circ{}_{298} = +159.7 \text{ kJ/mol} \qquad \text{R8.7}$$

There is a balance between the extent of combustion and gasification processes that is controlled by the products required and therefore the chosen stoichiometry and the reaction temperature and pressure. This is illustrated in Figure 8.3. By a change in conditions the product can include more CO and H_2, and this is achieved when power generation is the objective [Figure 8.3(a) and (b), respectively]. Figure 8.3(c) shows how the methane content can be the preferred product and is favored by low temperatures. Thus it can vary widely from process to process as shown in Figure 8.4.

Gasification of the char or carbon with carbon dioxide (Boudouard reaction) is usually the prime process together with the partial oxidation steps (R 8.2, R 8.4). The Boudouard reaction is endothermic and, for a given carbon in the absence of a catalyst, takes place several orders of magnitude slower than the C-O_2 reaction at the same temperature. The reaction proceeds very slowly at temperatures below 1000K, and is inhibited by its product, CO. If there is a significant amount of steam present, then reaction with carbon (R 8.6) takes place. This reaction has a high activation energy and the rate is proportional to the steam partial pressure.

These two reactions are endothermic, i.e., require heat in order to proceed. Therefore, the heat required for these desired reactions is supplied by the complete or partial combustion of a small proportion of the coal in oxygen or air. Other reactions taking place in the gasifier are as follows:

Hydrogenation:

$$C + 2H_2 \rightarrow CH_4 \qquad \Delta H^\circ{}_{298} = -88.4 \text{ kJ/mol} \qquad \text{R8.8}$$

Water shift reaction:

$$CO + H_2O \rightarrow CO_2 + H_2 \qquad \Delta H^\circ{}_{298} = -40.9 \text{ kJ/mol} \qquad \text{R8.9}$$

Methanation:

$$CO + 3H2 \rightarrow CH_4 + H_2O \qquad \Delta H^\circ{}_{298} = -206.3 \text{ kJ/mol} \qquad \text{R8.10}$$

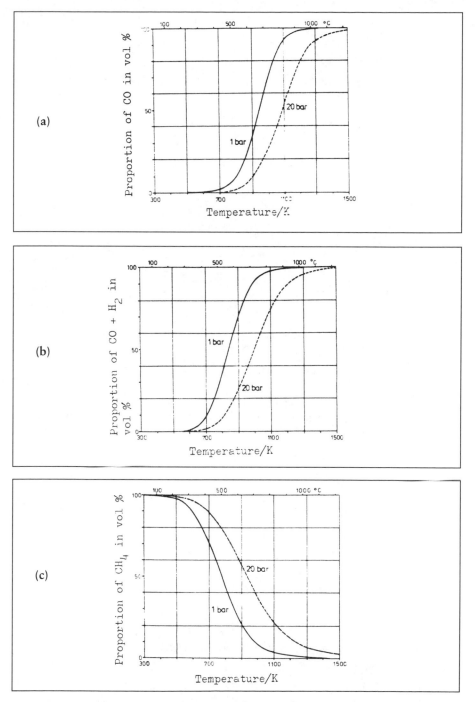

Figure 8.3 Equilibrium proportions of (a) CO, (b) CO + H$_2$, and (c) CH$_4$ in-product gas from reactions 8.2, 8.6, and 8.8, respectively (after Kristiansen, 1996).

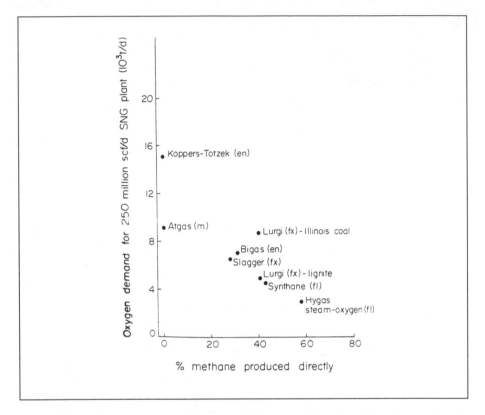

Figure 8.4 Variation of oxygen demand with % methane produced directly in coal gasification processes: fx, fixed bed; fl, fluidized bed; en, entrained flow; m, molten bath process.

Pyrolysis:

$$4C_n H_m \rightarrow mCH_4 + (4n - m)C \qquad \text{Exothermic} \qquad \qquad \text{R8.11}$$

Two gas-phase reactions are important for the final gas composition. The water-gas shift reaction has influence on the CO/H_2 ratio, which can be important if the gas is for use in synthesis.

Because of the hydrogen in the gas phase, hydrogenation of carbon also occurs, so that the gasifier gases also contain some methane. Methane is also produced by the reaction of carbon monoxide and hydrogen. The methanation reaction increases the calorific value of the gas, but is very slow except at high pressure and in catalytic-based reactions. The shift reaction increases the amount of hydrogen in the gases. The shift and methanation reactions are particularly important for substitute natural gas (SNG) production.

8.3
Gasification Methods

The process of gasification has been used for many years. Partial gasification, the formation of volatile components from coal, was observed in the seventeenth century and subsequently used in a practical sense to produce coal gas (town's gas) for gas lighting by Murdoch in 1797. This led to a major industry in many countries, the gas industry. However, complete gasification, in which the coke produced by the coal devolatilization step is also gasified, was not introduced until the mid-nineteenth century by Siemens and is widely used for industrial applications. Oxygen gasification was introduced in the mid-1920s as a means of producing a high-calorific-value town's gas.

The next phase in the development of gasification was to produce synthetic or substitute natural gas, the difference between the two definitions being that the first contains mainly methane, and the second, a methane/hydrogen mixture of acceptable (equivalent) flame characteristics. In parallel, the gasifier was developed for the needs of the petroleum industry. The emphasis was on the synthesis of gasoline (e.g., SASOL), of hydrogen for refinery purposes, and for synthetic, sulfur-free, diesel fuel, and for chemical feedstocks. These are now established techniques (Speight, 1993) but, for economic reasons, at the present time, the preferred feedstock is natural gas not coal.

The third strand is the development of the concept of power generation using gas turbines. The current preferred choice by industry is the use of high-efficiency gas turbines fired by natural gas in a CCGT mode or refinery-produced heavy residues. This was not always the case, and in the not too distant future may not be the most economic choice. There have been several attempts to fire coal directly in gas turbines, work initiated by the US Bureau of Mines and the Department of Minerals and Energy in Australia. There are considerable problems, as yet unsolved, with erosion and corrosion (Stringer and Meadowcroft, 1990). It was concluded that the use of gasified coal gas, either as a hot, completely combusted gas or as a flammable gas for combustion in situ is preferable to the direct use of coal, and this has led to the current interest in coal gasification processes, which are outlined below.

Early gasification technology has been set out in a number of excellent books (e.g., Schilling et al., 1979, Smoot, 1993, Kural, 1994). A vast number of processes have been proposed, but only those of contemporary interest are outlined. In addition, direct underground gasification of coal is a possibility but is a specialist field in its own right and is not considered here.

Three main systems are used: fixed-bed, fluidized-bed, and entrained-flow gasifiers, and molten baths have been proposed. Examples of these types will be discussed next.

Fixed-Bed Gasification Processes

Gasification processes based on fixed beds have been used since the start of the twentieth century for producing a lean gas using air gasification. Ash removal from fixed-grate systems was a problem, and this was solved by using rotating grates. However, in the current state of development most fixed-bed gasifiers are used with oxygen. The countercurrent flow enables good sensible heat usage in the bed with low product gas temperatures and high carbon conversion, and hence, high thermal efficiency. However, the bed uses large coal particles and the gasification intensity is lower than when pf is used as the feedstock. The main potential application of this type of technology is in the BGL gasifier shown in Figure 8.5. Generally, the fixed-bed dimensions are typically 4–5 m in depth and 4 m in diameter for a dry ash BGL (British Gas Lurgi) gasifier and are similar for the BGL slagging gasifier. The temperatures in steam/air dry ash gasifiers do not exceed the ash melting point. Steam/oxygen gasifiers may, however, be either dry ash or slagging, depending on the amount of steam blown into the gasifier in relation to the amount of oxygen. The melting characteristics of the mineral matter define the steam/oxygen ratio for the dry ash gasifier in order to withdraw the ash in dry form without too much slagging. The steam/oxygen ratio is significantly reduced for slagging operation in order to remove the ash in a liquid form. Depending on the melting characteristics of the mineral matter, a flux (for example, limestone) may be added to assist the melting for high refractory ashes and to control the viscosity of the slag.

The system can be considered to consist of four zones:

1. The top feeding zone
2. The main gasification zone
3. The combustion zone
4. The ash zone

These are analogous to top-feeding bed combustion set out in Chapter 5; the stoichiometry here, however, promotes gasified products. Typically, the products consist of H_2, 31 mol%, CO 57 mol%, CH_4 6 mol%, and CO_2 4.9 mol%,

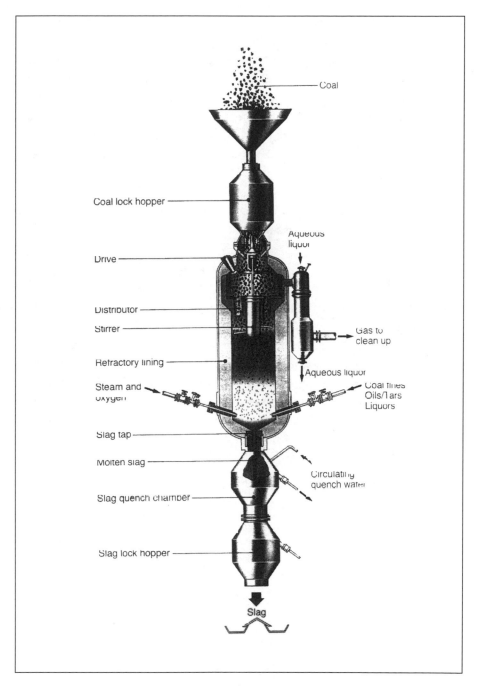

Figure 8.5 Outline of the BGL gasifier (BG plc).

but some tars and phenols are produced that have to be recycled. These figures are given in Table 8.3 for comparison with the products of other gasifiers.

The BGL process was developed for the production of synthetic natural gas. By adjustment of the feed, especially the amount of steam, the process can be adjusted to give products for use for power generation. It produces a high-calorific-value gas at a relatively low temperature, and gas-cleaning processes can readily be applied. The products can be cleaned using hot gas cleanup stages, or the relatively cool products can be cooled further to permit the use of conventional processes.

There are many rival processes (e.g., Schilling et al.,1979) that have been developed to a pilot stage but not fully commercialized, e.g., the Hygas process, the Cogas process, and the U-gas process.

Entrained-Flow Gasification Reaction

These processes were derived for the petroleum refining and chemical industry. The products need to have the correct proportion of CO/H_2 dependent on the application. Generally, equilibrium considerations indicate that as the temperature becomes higher, the amount of CO increases and the amount of methane decreases. Thus, as shown in Figure 8.4, the amount of CO_2 and H_2 resulting from the water-gas reaction increases as the temperature rises, and likewise from the Boudouard reaction. Thus, reactors operating at high temperature —

TABLE 8.3.
TYPICAL COMPOSITIONS OF GASES FROM GASIFIERS*

Gas composition (mol %)	Coal				Biomass
	Oxygen		Air		
	Fixed dry BGL	Shell, dry entrained	Texaco slurry	Fluidized bed	Oxygen entrained
CO	57	65	49	22	22
H_2	26–30	29	34	17	13–22
CO_2	4	2	10	7	20
CH_4	6	0.01	0.2	0.5	0.2
N_2	2	2	1	44	~1
NH_3 ppm	≡ fuel-N				0.2000

*sProducts depend on the coal used, the oxygen purity, and the reaction conditions.

with oxygen and high pressure — produce gases suitable for synthesis or for power generation.

A typical earlier plant is the Koppers-Totzek (KT) process (Figure 8.6), and a number (~60) of atmospheric pressure gasifiers have been built over a period of time for pf coal, oil, or gas usage. The reaction takes place at a maximum temperature of above 1500°C, the reaction time being about 1 second. The carbon is almost completely consumed, and the tars are gasified along with the carbon. The product gases leave at a temperature of about 1500°C, and their cooling to 300°C heats the steam for the process. More than half of the ash is removed as molten slag at the bottom of the burner. These can be used at higher pressures. Typically, the gas consists of 55–66% CO, 21–32% H_2, 0.1% CH_4, and 7–12% CO_2. The sulfur and nitrogen in the coal are converted to H_2S and NH_3, respectively, and have to be removed from the gas stream. While these can be removed by conventional scrubbing techniques, there are some attractions as far as the thermodynamics of the IGCC system is concerned to remove the trace species at elevated temperatures (Clift and Seville, 1993, Mitchell, 1998).

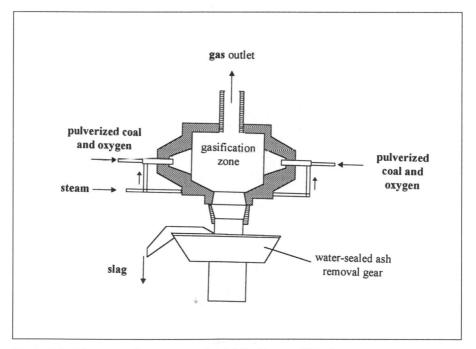

Figure 8.6   Diagrammatic representation of the Koppers-Totzek process.

In entrained-flow gasifiers, the fine coal particles can be fed either dry (normally using nitrogen as transport gas) or wet (carried in a water slurry). Some gasifiers use two-stage feeding to improve the thermal efficiency and reduce both the sensible heat in the raw gas and the oxidant requirements. Depending upon the method of coal feeding, dry or wet slurry, the entrained-flow gasifiers can accept almost any type of coal.

Essentially, all entrained-flow gasifiers use oxygen as the oxidant and operate at high temperatures well above ash slagging conditions, to ensure high carbon conversion. Of the three main gasifier types, the entrained-flow gasifiers operate at the highest temperature. To obtain the high operation temperature, the entrained-flow gasifier has a high oxygen requirement. Negligible amounts of methane are produced, and there are few tars and heavy hydrocarbons, if any, in the product gas.

Typically, such gasifiers work at pressures up to 35 bars. There are a number of oxygen-based processes based on this technique at the present time, namely:

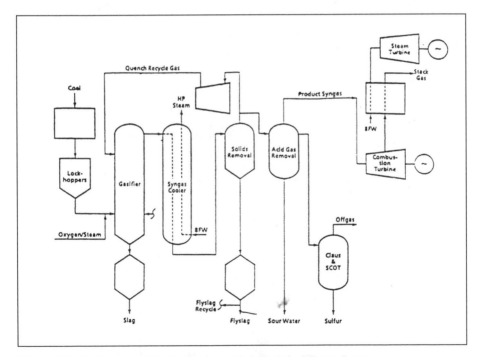

Figure 8.7   Shell coal gasification process (SCGP) typical flow scheme.

The Shell gasification process, which is shown in outline in Figure 8.7 and has formed the basis of a 250-MWe demonstration power plant in Buggenum in the Netherlands since 1998.

The Prenflo process (Krupp Koopers), which has been demonstrated since 1996 in Puertollano in Spain and that has many characteristics similar to the Shell system. It is a 335-MW IGCC power plant using coal mixed with petroleum coke from a refinery.

The Texaco process, using a coal-water slurry injection system (Figure 8.8) that has been in commercial operation in the US for a number of years.

The Destec entrained-flow gasifier, which has been developed in the US and is a two-stage gasifier. This is shown in Figure 8.9.

The Japanese air-blown entrained gasifier developed in the early 1990s. Much work has been undertaken on this, but the pilot plant is now subject to feasibility studies.

Fluidized-Bed Processes

The fluidized bed is operated at a constant temperature, usually below the ash fusion temperature, thereby avoiding agglomeration and clinker formation and defluidizes the bed. As coal particles are consumed or fragmented during gasification, the smaller particles are entrained with the hot raw gas as it leaves the reactor; these char particles are recovered and recycled to the reactor.

Fluidized-bed gasifiers may differ in ash conditions, being run either dry or agglomerated. The agglomerated ash operation improves the ability of the process to gasify high-rank coals efficiently. Conventional dry ash operation has traditionally operated on low-rank coals.

The immediate forerunner is the original Winkler process, which is a fluidized-bed system that uses steam and air or oxygen at atmospheric pressure. A typical plant outline is shown in Figure 8.10.

As the drive toward higher efficiency operation for electricity generation has developed, particular attention has been directed to air-blown fluidized-bed units. In particular, two units are attracting attention:

1. The Pinon Pine IGCC Demonstration Plant in Reno in the US. This 100-MWe plant uses a pressurized fluidized bed (KRW) with in-bed and high-temperature sulfur removal and high-temperature ceramic filters (99% removal).

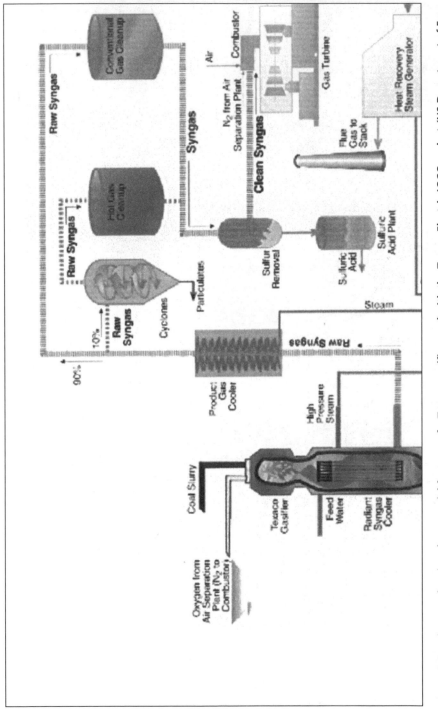

Figure 8.8 Diagram showing the essential features of a Texaco gasifier used with the Tampa Electric IGCC project (US Department of Energy/Tampa Electric Company, Topical Report Number 6, October 1996).

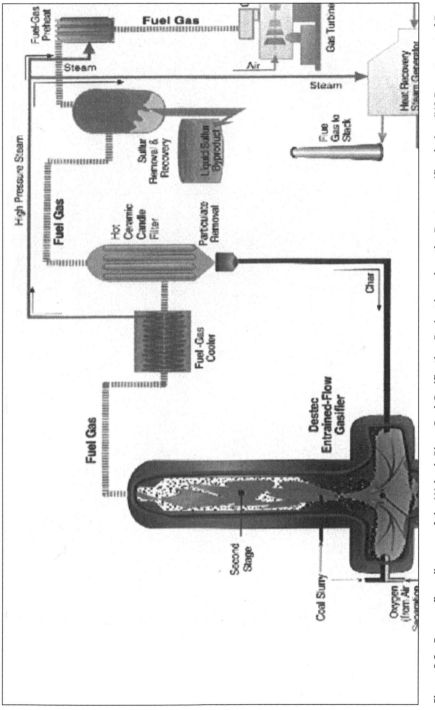

Figure 8.9 Process flow diagram of the Wabash River Coal Gasification Project based on the Destec gasifier design (US Department of Energy/Wabash River Coal Gasification Project, Topical Report Number 7, November 1996).

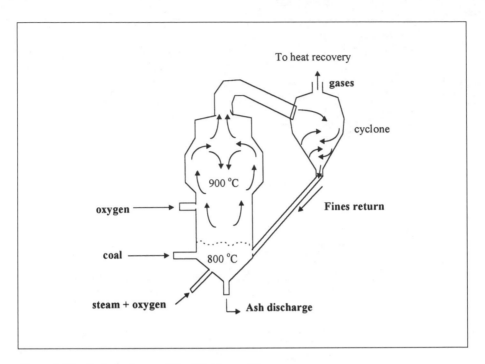

Figure 8.10   Basic features of the Winkler gasifier.

2.   The Air-Blown Gasifier (ABGC) developed in the UK and formerly
     termed the Topping Cycle, shown in Figure 8.11. This involves an air-
     blown partial (70–80%) gasifier with combustion of the char in a sepa-
     rate combustor. The gasified gas has hot-gas cleanup associated with it.
     Steam produced by heat recovers from the gas turbine combine with
     steam from the circulating fluidized-bed combustion of the char resulting
     in a high-efficiency (> 50%) system. It is currently undergoing feasibility
     studies in the UK.

8.4
Detailed Reaction Mechanisms and Intrinsic Kinetics

Gasifiers are fed with oxygen or air to generate the heat required for the gasification processes by the partial combustion of a part of the coal. In general, the larger
the oxygen to coal ratio, the higher the temperature and the easier complete con-

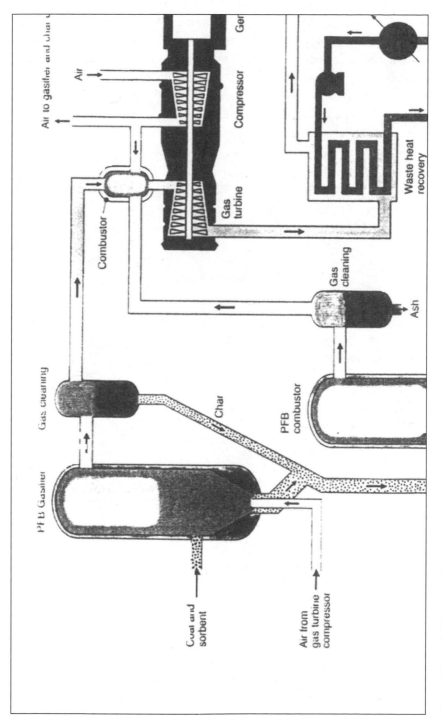

Figure 8.11 Air-blown gasifier combined cycle (ABGC).

version is attained. However, this will also lead to production of gas high in CO_2 and H_2O, which in most cases are unwanted. By lowering the amount of O_2, it requires increasingly more conversion through reaction with less reactive CO_2 and H_2O to attain a similar burn-out performance, but at a certain oxygen level coal conversion will become incomplete. Therefore, careful optimization of gasification processes, in terms of operability, requires a good understanding of coal reactivity and kinetics of the CO_2 and H_2O reactions, and these have been extensively studied, particularly over the last 20 years. There have been a number of significant reviews of the mechanisms, e.g., Laurendeau (1978), Van Heek and Mühlen (1991), Smith et al. (1994), Kristiansen (1996), Takematsu and Maude (1991). This following analysis takes the approach of the last authors.

A general kinetic expression for the overall reaction rate can be written as

$$\frac{dX}{dt} = r_s\left(p_i t_p\right) f(X) \qquad\qquad E\ 8.1$$

where $r_s\ (p_i T_p)$ is the intrinsic surface-reaction rate depending on the partial pressure of the component i, p_i, and the particle temperature, T_p, and $f(X)$ is a factor that describes the total number of reactive sites per unit volume of char as a function of the carbon conversion (X).

Reaction Mechanism and Intrinsic Reaction Kinetics of the Water-Gas Reaction

A number of models of gasification have been proposed. In order to describe gasification with steam, four mechanistic models have been proposed that are based on the following surface reactions:

Step 1:	$C_{fas} + H_2O \leftrightarrow C(O) + H_2$	R 8.12
Step 2:	$C(O) \rightarrow C_{fas} + CO$	R 8.13
Step 3:	$C_{fas} + H_2 \leftrightarrow C(H)_2$	R 8.14
Step 4:	$C_{fas} + \frac{1}{2}H_2 \leftrightarrow C(H)$	R 8.15

Here C_{fas}, etc., represent free-active sites (cf. R 4.6). In all the models the first step (reaction R 8.12) is the dissociation of a water molecule at a carbon active site, releasing hydrogen and forming an oxidized surface complex. The actual gasification of carbon occurs in step 2 (reaction R 8.13), in which the carbon-oxygen complex subsequently produces a molecule of CO and a new active site. The rate-limiting step in water vapor gasification is probably the desorption of the carbon-oxygen surface complex (reaction R 8.12). The activation energies found for the desorption step are around 250–295 kJ/mol.

The models differ in the way they account for the inhibition step caused by the formation of hydrogen. The oxygen-exchange model is based on the first and second reactions, whereby reaction R 8.12 is assumed to be reversible. According to this model, hydrogen inhibition is attributed to the equilibrium of the dissociation reaction (reaction R 8.12). The first, more popular, hydrogen-inhibition model is composed of the first, second, and third reactions, but in this case reaction R 8.12 is assumed to be irreversible. Hydrogen inhibition is attributed to the formation of the $C(H)_2$-complex (reaction R 8.14). In the second hydrogen-inhibition model the third reaction is replaced by the fourth reaction, which describes dissociative chemisorption of hydrogen (reaction R 8.15). The complete model is composed of reaction R 8.12, which also is irreversible, and reactions R 8.13 and R 8.15. Recently, a combined oxygen-exchange/H_2-adsorption model was proposed, and this consists of the first, second, and third reactions. In this, the hydrogen inhibition is described by both the reverse of the dissociation reaction (reaction R 8.12) and the reversible H_2 adsorption (reaction R 8.14).

All the models yield basically the identical intrinsic surface-rate equation for r_s:

$$r_s = \frac{k_{12}\, p_{H_2O}}{1 + \dfrac{k_{12}}{k_{13}}\, p_{H_2O} + f\!\left(p_{H_2}\right)} \qquad\qquad \text{E 8.2}$$

where k_i is the rate constant, $k_i = k_0 \exp(-E_i/RT)$, E_i is the activation energy of the reaction, T is the temperature, and p_i is the partial pressure. Only the term in the denominator describing the function of the partial pressure of H_2 changes with the models. These are:

1. Oxygen-exchange: $\quad f\!\left(p_{H_2}\right) = \dfrac{k_{-12}}{k_{13}}\, p_{H_2}$

2. Hydrogen inhibition by formation of a $C(H)_2$-complex: $\quad f\!\left(p_{H_2}\right) = \dfrac{k_{14}}{k_{-14}}\, p_{H_2}$

3. Hydrogen inhibition by formation of a C(H)-complex: $f\left(p_{H_2}\right) = \dfrac{k_{15}}{k_{-15}} p_H^{0.5}$

4. Hydrogen inhibition by oxygen-exchange and formation

of a C(H)$_2$-complex: $f\left(p_{H_2}\right) = \left(k_{14} + \dfrac{k_{-12}}{k_{13}}\right) p_{H_2} + k_{14} \dfrac{k_{-12}}{k_{13}} p_{H_2^2}$

The rate expressions from the different models enable discrimination between these because of their different order of dependence on the hydrogen partial pressure.

It can be shown that during gasification in steam, the reaction rate almost linearly decreases with increasing carbon conversion. This means that the reaction is approximately first order in carbon, which means that the reactivity of the char is almost constant during gasification. The presence of H_2 lowers the initial rate of gasification, and results in a gradual decrease in reactivity of the char with increasing carbon conversion. Furthermore, it was found that the decrease in reactivity is more pronounced the lower the temperature, and the higher the hydrogen partial pressure.

The initial reaction rates have been tested on three different Langmuir-Hinshelwood/Hougen-Watson—type rate expressions that differ only in the way H_2 inhibition is accounted for. The expressions contain H_2-inhibition terms with a first-order dependence on pH_2, a square root dependence on PH_2, and a sum of a first-order and a second-order dependence on PH_2. The hydrogen inhibition seems to be caused by reaction of the carbon active sites with H_2 producing a C(H)$_2$-complex. The following rate expression was obtained by Weeda et al. (1993a):

$$r_s = \frac{32.8 \exp(-209,200 / RT) p_{H_2O}}{1 + 4.9\ 10^{-5} \exp(-24,500 / RT) p_{H_2O} + 11.10^{-7} \exp(114,200 / RT)\sqrt{p_{H_2}}}$$

$$\text{E 8.3}$$

where r_s is in g/s, p_i is in Pa, T is in K, and $R = 8.3$ J/ mol/K.

Recently, Weeda (see Kristiansen, 1996) reported his latest results on steam gasification inhibited by hydrogen. He tried to model the decrease in reactivity during gasification as a decrease in the number of reactive sites per unit char-surface area. Again he found the data to scale best with the square root of p_{H_2}. The same values for the rate constants and activation energies in the kinetic rate equation were found. A positive value was obtained for the apparent activation energy E_{-1}, an indication that the value of the apparent rate constant k_{-1} increases as the temperature increases. Since $E_{-1} = E_1 - E_2$, a positive value for E_{-1} suggests that the activation energy for the formation of a C(O)-complex is greater than that for its decomposition. This implies that with increasing temperature, the rate of C(O) formation increases faster than the rate of C(O) decomposition. Hence, the rate of gasification will become independent of at sufficiently high pressures and temperatures, and depend only on the step in which the oxidized complex dissociates (equation E 8.2). Conversely, the rate of C(O) formation is the rate-limiting step at sufficiently low pressures and temperatures. A unified theory of the mechanism has been postulated by Moulijn and Kapteijn (1995).

Reaction Mechanism and Intrinsic Kinetics of the Boudouard Reaction

Several mechanisms have been suggested for the reaction of carbon with carbon dioxide, and have been reviewed by a number of people (Laurendeau, 1978, Moulijn and Kapteijn, 1995). The proposed schemes involve either a reversible oxygen-exchange mechanism (Ergum and Tiensuu, 1959, Ergun, 1962) or a unidirectional exchange of oxygen at the surface coupled with chemisorption of CO upon active sites proposed by Gadsby et al. (1948). The mechanisms are based on the following surface reactions:

Step 1: $\qquad C_{fas} + CO_2 \leftrightarrow C(O) + CO \qquad$ R 8.16

Step 2: $\qquad C(O) \rightarrow CO + C_{fas} \qquad$ R 8.17

Step 3: $\qquad C_{fas} + CO \leftrightarrow C(CO) \qquad$ R 8.18

In all models the first step is the dissociation of a carbon dioxide molecule at a carbon active site (R 8.16), releasing carbon monoxide and forming an oxidized surface complex. The actual gasification step of carbon occurs in the second step (R 8.17), in which the carbon-oxygen complex subsequently produces a molecule of CO and a new active site. As with steam gasification, the rate-limiting step in carbon dioxide gasification is the desorption of the carbon-oxygen surface complex (R 8.18).

The mechanism proposed by Ergun (1962) involves the first and second reactions. The reverse of the second reaction (R 8.17) is relatively slow compared to the forward and reverse rates of the first reaction (R 8.16). The rate for the above mechanism can be described by the Langmuir-Hinshelwood rate equation, indicating that the C/CO_2 reaction rate is dependent on CO and CO_2 partial pressures and is inhibited by the presence of CO.

An alternative mechanism was proposed by Gadsby et al. (1948). This mechanism involves all three reactions. In this model the hydrogen inhibition is ascribed to the formation of the C(CO)-complex (R 8.17).

Hampartsoumian et al. (1993) studied the reaction of chars from three UK coals with carbon dioxide to determine which of the two mechanisms controls the reaction between char and carbon dioxide. On the basis of the calculated activation energies for k_{16}, k_{-16}, and k_{17}, it was suggested that the oxygen-exchange mechanism holds. The activation energies obtained for the reverse of the first reaction (R 8.16), which represents the removal of a C(O) site, are all positive, thus providing clear evidence of the reversibility of the carbon-CO_2 reaction and the inhibiting effect of CO.

The rate for the oxygen-exchange mechanism can be described by the Langmuir-Hinshelwood rate equation:

$$R_a = \frac{k_{16}P_{CO_2}}{1 + K_a P_{CO} + K_b P_{CO_2}} \qquad\qquad \text{E 8.4}$$

where $K_a = k_{-16}/k_{17}$, $K_b = k_{16}/k_{17}$, and R_a (g/s) is the rate of consumption of carbon due to the two reactions in the oxygen-exchange mechanism. Hampartsoumian et al. (1993) obtained values for k_{16}, K_a, and K_b in the temperature range 900–1300K. The effect of temperature on the rate constants within this range can be summarized as: k_{16} increases with increasing temperature, K_a decreases with increasing temperature, and K_b decreases with increasing temperature, which indicates that the retardation effect of CO on the carbon-CO_2 reaction is reduced with increasing temperature.

There is a wide variation in rate constants and activation energies, perhaps not surprisingly, since types of carbon, experimental systems, and methods employed in analyzing the kinetic data varied. All authors predicted a negative value for E_a. Many authors obtained values of E_{k16} within 170–220 kJ/mol and E_{k-18} within 190–240 kJ/mol. However, Strange and Walker (1976) obtained a value of 414.5 for E_{k16} and 364.2 for E_{k-18} by using pure natural graphite, and suggested that impurities in the carbon samples of others could have caused the lowering of the values of E_{k16} and E_{k-18}. The work of Wu et al. (1988) and Gadsby et al. (1948) yielded negative values of E_{k-16}, but this is not possible if the oxygen-exchange mechanism holds. The value of -70.4 kJ/mol by Gadsby et al. (1948) can possibly be due to experimental error [$E_{k-16} = E_{16} + E_a - E_b$ and E_a and E_b are large and of opposite sign).

The results recently reported by Weeda (see Kristiansen, 1996) give a negative value for Ea, meaning that the activation energy for the decomposition of the C(O)-complex is larger than for its formation. In this case, the decomposition of the C(O)-complex controls the rate of gasification at low temperatures. With increasing temperature, the rate of the process with the highest activation energy increases most rapidly, and as a consequence, the rate of C(O) formation decreases relative to the rate of C(O) decomposition. In other words, the reaction order in increases at increasing temperature. The activation energy of K_b was found to be negative (Hampartsoumian et al., 1993). Consequently, inhibition by CO decreases with increasing temperature, as with increasing temperature the rate of decomposition of the C(O)-complex grows faster than that of its decomplexation by CO. It is clear that some of these processes are very complex (Molina and Mondragon, 1998).

8.5
Process Problems and Environmental Considerations

The efficient use of coal or reduction of CO_2 emissions has been the driving force in developing gasification technologies. They have been developed to provide the energy industry with environmentally acceptable and economically competitive alternatives to conventional power generation technologies.

Low-CV gas cannot be used for natural gas replacement or for synthesis gas production because of the high content of nitrogen. Low-CV gas can be used as only a fuel gas at the present time; this is in gas turbines in a thermodynamically efficient system. It is used in an integrated system.

Medium-calorific-value gas can be used for natural gas replacement or for synthesis gas production since natural gas is plentiful at the moment in substitute natural gas except for strategic reasons. In synthesis gas production the gas can be further upgraded by changing the carbon monoxide/hydrogen ratio catalytically to obtain a gas mixture that can be used as a feedstock for a variety of processes such as Fischer-Tropsch synthesis, methanol, acetic acid, and ammonia production.

Both low-CV and medium-CV gas can be used in gas-steam combined power cycles, which is a part of the integrated gasification combined cycle (IGCC) system. In this process, fuel gas is generated in a coal gasifier and subsequently purified before being burned in the gas turbine. In the turbine, where it is burned with compressed air to provide a stream of hot, high-pressure gas that drives the turbine, to generate electricity, this gas stream must be clean enough for the gas turbine, and this necessitates gas filtration and the removal of any corrosive components. Heat from the gasifier and from the gas turbine exhaust raises steam to power a steam turbine, which generates additional electricity. Their use requires an optimized thermodynamic cycle with attention to "pinch points" and the minimization of the use of auxiliary power units and other energy losses. If air separation is used, the high-pressure nitrogen product gas has to be used in conjunction with the gas turbine so that the energy used in compression is recovered.

The problems encountered in the gasification of coal are discussed next.

Gas Cleanup

Depending upon the gasifier, the gases may contain tar, ash particles, ammonia (NH_3), hydrogen cyanide (HCN), and acid gases (H_2S, COS) and some aromatics. Whatever the application of the gasifier, power generation, SNG production, coal liquefaction, these impurities have to be reduced, but the manner of doing so depends on the application.

Tar if produced is removed by a tar trap. Ash is removed by cyclones and filtration. Ammonia and the acid gases can be produced, and it is possible to use wet scrubbing methods for syngas applications because the gases have to be cooled anyway. For power generation, hot-gas cleanup is the most desirable method, giving a small gain in efficiency, and hot-gas filtration is now a fairly developed technology and units are readily available. H_2S and CO_3 can be removed in bed techniques using limestone. The reactions are

$$CaCO_3 \rightarrow CaO + CO_2 \qquad \text{R 8.19}$$
$$CaO + H_2S \rightarrow CaS + H_2O \qquad \text{R 8.20}$$

Also possible is the use of the metal oxide (ZnO, NiO) sorbents:

$$MO + H_2S \rightarrow MS + H_2O \qquad \text{R 8.21}$$
$$MO + COS \rightarrow MS + CO_2 \qquad \text{R 8.22}$$

The sorbent can be regenerated to some extent.

Ammonia is a particular problem because it is converted into NO_x in the gas turbine, and emission limits could be exceeded. The amount in the gas is dependent on the fuel-N content in the coal, but is commonly about 1000 ppm. It can be removed at lower temperatures, but there are no really satisfactory hot-gas removal techniques at present. There are catalytic methods available, and NO can be added to react with the NH_3; none give high levels of removal.

Gas Turbine Requirements

The gas turbine itself imposes a number of restrictions on the gas composition. The gas can never be "pure" and will contain some particulate material, ammonia, and acid gas. The particulate is defined by its loading and maximum particle size. If ammonia is present, a fuel-N component, then a two-stage low-NO_x burner must be used. The choice of materials determines the upper acid composition.

It is clear that many of these problems can be overcome by clever gasifier design, but it is also clear from Section 8.4 that a detailed understanding of the reaction in a gasifier is far from clear. For design purposes empirial equations have to be used; the detailed modeling of the individual reactions in a gasifier in a way similar to that used for gas-phase combustion processes is not possible at present. However, there is very considerable scope for advances in the future in both gasifier and gas turbine design that will lead to higher system efficiences.

REFERENCES

Aarna, I. and Suuberg, E.M. (1997) *Fuel*, **76**, 475.

Aarna, I. and Suuberg, E.M. (1998) *Twenty-seventh International Symposium on Combustion*, The Combustion Institute, Pittsburgh, 2933.

Abbas, A.S. and Lockwood, F.C. (1985) *J. Inst Energy*, **58**, 112.

Abbas, T., Costen, P.G. and Lockwood, F.C. (1996) *Twenty-sixth International Symposium on Combustion*, The Combustion Institute, Pittsburgh, 3041.

Arthur, J.R. (1951) *Trans. Faraday Soc.*, **47**, 164.

Attar, A. and Hendrickson, G.G. (1992) Coal Structure (Ed. Meyers, R.A.), Academic Press, New York, 131–197.

Azhakesan, M., Bartle, K., Murdoch, P., Taylor, J. and Williams, A. (1991) *Fuel*, **70**, 322.

Badin, E.J. (1984) Coal Combustion Chemistry-Correlation Aspects, *Coal Sci. and Tech.*, **6**, Elsevier, Amsterdam.

Badzioch, S. and Hawksley, P. (1970) *Ind. Engg. Chem. Res.*, **9**, 1137.

Barret, R.E. (1990) *Power*, **134** (2), 41.

Battcock, W.V. and Pillai, K.K. (1977) *Proc. Fifth International Conference on Fluidised Bed Combustion*, Mitre Corp., Washington D.C.

Baum, M.M. and Street, P.J. (1971) *Comb. Sci. & Tech.*, **3**, 231.

Bayliss, D.J., Schroeder, A.R., Johnson, D.C., Peters, J.E., Krier, H. and Buckins, R.O. (1994) *Comb. Sci. & Tech.*, **98**, 185.

Beeley, T., Crelling, Gibbins, J., Hurt, R., Lunden, M., Man, C., Williamson, J. and Yang, N. (1996) *Twenty-sixth International Symposium on Combustion*, The Combustion Institute, Pittsburgh, 3103.

Beer, J.M. (1996) *Comb. Sci. and Tech.*, **121** 169.

Bemtgen, J.M., Hein, K.R.G. and Minchener, A.J. (1995) Clean Coal Technology Programmes, 1992–1994, *Vol. 11, Combined Combustion of Biomass Sewage Sludge and Coals*, European Commission, DG-XII.

BP Statistical Review of World Energy (1998) The British Petroleum Company plc, London. http:www.bpamoco.com/worldHenergy

Braide, K.M., Isle, G.l., Jorden, J.B. and Williams, A. (1979) *J. Inst. Energy*, **52**, 115.

Bray, K.N.C. (1996) *Twenty-sixth International Symposium on Combustion*, The Combustion Institute, Pittsburgh, 1.

Brown, A.L. and Fletcher, T.H. (1998) *Energy and Fuels*, 12, 745.

Burchill, P. and Welch, S.L. (1989) *Fuel*, 68, 100.

Carpenter, A.M. (1988) Coal Classification. *IEACR/12*, IEA Coal Research, London.

Carpenter, A.M. (1998) Switching to Cheaper Coals for Power Generation, IEA Coal Research, London.

Carpenter, A.M. and Skorupska, N.M. (1993) Coal Combustion–Analysis and Testing, *IEACR/64*, IEA Coal Research, London.

Chagger, H.K, Goddard, P., Murdoch, P., Williams, A. (1991) *Fuel*, 70, 1137.

Chan, L.K., Sarofim, A.F., Beer, J.M. (1983) *Comb. and Flame*, 52, 37.

Charpeney, S., Serio, M.A. and Solomon, P.R. (1992) *Twenty-fourth International Symposium on Combustion*, The Combustion Institute, Pittsburgh, 1189.

Chen, H.R. and Driscoll, J.F. (1991) *Twenty-third International Symposium on Combustion*, The Combustion Institute, Pittsburgh, 281.

Chen, J-Y., Mann, A.P. and Kent, J.H. (1992) *Twenty-fourth International Symposium on Combustion*, The Combustion Institute, Pittsburgh, 1381.

Chen, S., Heap, M., Pershing, D., Martin, G. (1982) *Nineteenth International Symposium on Combustion*, The Combustion Institute, Pittsburgh, 1271.

Chen, Y., Charpeney, S., Jensen, A., Wojitowicz, M.A. and Serio, M.A. (1998) *Twenty-seventh International Symposium on Combustion*, The Combustion Institute, Pittsburgh, 1327.

Clarke, A.G., Pourkashanian, M. and Williams, A. (1987) *First European Dry Fire Coal Conference*, The Institute of Energy, London.

Clarke, L.B. and Sloss, L.L. (1992) Trace Elements–Emissions from Coal Combustion and Gasification, *IEACR/49*, IEA Coal Research, London.

Clift, R. and Seville, J.P.K. (1993) Gas Cleaning at High Temperature, Blackie Academic, London.

Couch, G. (1994) Understanding Slagging and Fouling During pf Combustion, *IEACR/72*, IEA Coal Research, London.

Davidson, R.M. (1994) *Fuel*, 73, 988.

Davidson, R.M. (1997) Co-processing Waste with Coal, *IEAPER/36*, IEA Coal Research, London.

Doing, A. and Morrison, G. (1997) The Use of Natural Gas in Coal Fired Boilers, *IEAPER/35*, IEA Coal Research, London.

Ergun, S. (1962) US Bureau of Mines Bulletin, 598.

Ergun, S. and Tiensuu, V.H. (1959) *Acta Crystallogr.*, 12, 1050.

Essenhigh, R.H. (1967) *Ind. Engg. Chem.*, 59, 52.

Essenhigh, R.H. and Mescher, A.M. (1996) *Twenty-sixth International Symposium on*

Combustion, The Combustion Institute, Pittsburgh, 3085.

Fairweather, M., Jones, W.P. and Lindstedt, R.P. (1992) *Comb. and Flame*, **89**, 45.

Field, M.A., Gill, D.W., Morgan, B.B. and Hawksley, P.G.W. (1967) The Combustion of Pulverised Coal, The British Coal Utilisation Research Association, Leatherhead.

Fiveland, W. and Latham, C.E. (1997) *Combustion Technologies for a Clean Environment*, Gordon and Breach, New York, 111.

Fiveland, W. and Wessel, R. (1991) *J. Inst. Energy*, **64**, 41.

Fletcher, T.H., Solum, M.S., Grant, D.M., Critchfield, S. and Pugmire, R.J. (1990) "Solid State ^{13}C and ^{1}H NMR Studies Of The Evolution Of The Chemical Structure Of Coal Char And Tar During Devolatilization," *Sandia Report No. SAND89-8793*.

Frost, D.C., Wallbank, B. and Leeder, W.R. (1978) Analytical Methods for Coal and Coal Products (Ed. Karr, C. Jr.) Academic Press, New York 349.

Gadsby, J., Long, F.J., Sleightholm, P. and Sykes, K.W. (1948) *Proc. of the Roy. Soc.*, A193, 357.

Gerstein, B.C., Dubois Murphy, P. and Ryan, L.M. (1982) "Aromaticity in coal" in Coal Structure, (Ed. Meyers, R.A.) Academic Press, New York.

Given, P.H. (1960) *Fuel*, **39**, 147.

Glarborg, P., Alzueta, M.U., Dam-Johansen, K. and Miller, J.A. (1998) *Comb. and Flame*, **115**, 1.

Glass, J. and Wendt, J. (1982) *Nineteenth International Symposium on Combustion*, The Combustion Institute, Pittsburgh, 1243.

Grondhovd, G.H., Sondreal, E.A., Kotowski, J. and Wiltsee, G. (1982) Low Rank Coal Technology, Noyes Data Corporation, NJ, US.

Hampartsoumian, E., Pourkashanian, M. and Williams, A. (1989) *J. Inst. Energy*, **62**, 48.

Hampartsoumian, E., Pourkashanian, M. and Williams, A. (1993) *Comb. Sci. & Tech.*, **92**, 105.

Hargrave, G., Pourkashanian, M. and Williams, A. (1986) *Twenty-first International Symposium on Combustion*, The Combustion Institute, Pittsburgh, 221.

Hill, S.C., Smoot, L.D. and Smith, P.J. (1984) *Twentieth International Symposium on Combustion*, The Combustion Institute, Pittsburgh, 1391.

Howard, J.R. (Ed.) (1983) Fluidised Beds, Combustion and Applications, *Applied Science*, London.

Hurt, R.H. and Davis, K.A. (1994) *Twenty-fifth International Symposium on Combustion*, The Combustion Institute, Pittsburgh, 561.

Hurt, R.H., Lunden, M.M., Brehob, E.G. and Maloney, D.J. (1996) *Twenty-sixth International Symposium on Combustion*, The Combustion Institute, Pittsburgh, 3169.

IEA Coal Industry Advisory Board (1995) Industry Attitudes to Steam Cycle Clean Coal Technologies, *OECD/IEA*.

Ishida, M. and Wen, C.Y. (1968) *AIChE*, **14**, 311.

JANAF Thermochemical Tables Database (1995) National Institute of Standards and Technology, Gaithersburg, MD, US.

Jones, J.M., Patterson, P.M., Pourkashanian, M., Rowlands, L. and Williams, A. (1997) *Fourteenth Annual Int. Pittsburgh Coal Conference*, China.

Jones, J.M., Pourkashanian, M., Williams, A., Chakraborty, R.K. and Sykes, J. (1998) *Fifteenth Annual Int. Pittsburgh Coal Conference*, USA.

Kelemen, S.R., Gorbaty, M.L., Vaughn, S.N. and George, G. (1991) *Am. Chem. Soc. Div. Fuel Chem.*, **36**, No. 1225.

Khan, I.M. and Greeves, G. (1974) in Heat Transfer and Flames (Eds. Afgan, N.H. and Beer, J.M.), Scripta Books Co., Washington, D.C., 389.

Kobayashi, H., Howard, J.B. and Sarofim, A.F. (1976) *Sixteenth International Symposium on Combustion*, The Combustion Institute, Pittsburgh, 411.

Kristiansen, A. (1996) *Understanding Coal Gasification*, IEACR/86, IEA Coal Research, London.

Kulasekavan, S. and Agarwal, P.L. (1998) *Fuel*, **77**, 1033.

Kunii, D. and Levenspiel, O. (1991) Fluidization Engineering, 2nd Edition, Butterworth-Heinemann, Boston.

Kuo, K.K. (1986) Principles of Combustion, John Wiley and Sons, New York.

Kural, O. (1994) Coal. Istanbul Technical University, Istanbul.

Ladner, W.R. (1978) *J. Inst. Fuel*, **51**, 67.

Laurendeau, N.M. (1978) *Prog. Energy Comb. Sci.*, **4**, 221.

Lockwood, F., Rizvi, S., Lee, G. and Whaley, H. (1984) *Twentieth International Symposium on Combustion*, The Combustion Institute, Pittsburgh, 513.

Lockwood, F. and Romo-Millares, C.A. (1992) *J. Inst. Energy*, **65**, 144.

Ma, H.K. and Wu, F.S. (1992) *Int. Comm. Heat Mass Transfer*, **19**, 409

MacNeil, S. and Basu, P. (1998) *Fuel*, **77**, 269.

Macrae, J.C. (1966) An Introduction to the Study of Fuel, Elsevier, Amsterdam.

Magnusson, B.F. and Hjertager, B.H. (1977) *Sixteenth International Symposium on Combustion*, The Combustion Institute, Pittsburgh, 719.

Marsh, H. (1989) Introduction to Carbon Science, Butterworths, London.

Mason, H.B., Waterland, L.R., Chan, I.S. and Drennard, S.C. (1998) *Paper IPP41*, Int. Gas Research Conference, GRI, Chicago.

McHale, E.T. (1985) Review of CWF, *Comb. Tech. Energy Prog.*, **5**, 15.

Merrick, D. (1984) Coal Combustion Conversion Technology, Macmillan, London.

Miller, J. and Bowman, C.J. (1989) *Prog. Energy Comb. Sci.*, **15**, 287.

Missaghi, M., Pourkashanian, M., Williams, A. and Yap, L. (1991) "The prediction of

NO emissions from an industrial burner." *Proc. American Flame Day Conference*, San Francisco.

Mitchell, R.E. and Hurt, R. (1992) *Twenty-fourth International Symposium on Combustion*, The Combustion Institute, Pittsburgh, 1243.

Mitchell, S.C. (1998) Hot Gas Cleanup of Sulphur, Nitrogen, Minor and Trace Elements, *CCC/12*, IEA Coal Research, London.

Molina, A. and Mondragon, F. (1998) *Fuel*, 77, 1831.

Moss, J.B., Stewart, C.D. and Syed, K.J. (1988) *Twenty-second International Symposium on Combustion*, The Combustion Institute, Pittsburgh, 413.

Moulijn, J.A. and Kapteijn, F. (1995) *Carbon*, 33, 1155.

Muller, C.H., Schofield, K., Steinburg, M. and Broida, H.P. (1979) *Seventeenth International Symposium on Combustion*, The Combustion Institute, Pittsburgh, 867.

Mullins, O.C., Mitra-Kirtley, S., Van Elp, J. and Cramer, S.P. (1993) *Appl. Spectrosc.*, 47, 1268.

Murdoch, P.L., Pourkashanian, M. and Williams, A. (1984) *Twentieth International Symposium on Combustion*, The Combustion Institute, Pittsburgh, 1409.

Nasserzadeh, V., Swithenbank, J., Lawrence, D., Garrod, N.P., Silvennoinen, A. and Jones, B. (1993) *J. Inst. Energy*, 66, 169.

Niksa, S. (1993) *Twenty-fifth International Symposium on Combustion*, The Combustion Institute, Pittsburgh, 537.

Niksa, S. (1996) Coal Combustion Modelling, *IEAPER/31*, IEA Coal Research, London.

Niksa, S., Stallings, J., Mehta, A., Kornfield, A., Hurt, R., Mugio, L. (1999) EPRI-DOE-EPA Combined Utility Air Pollution Control Symposium—The MEGA Symposium, vol. 2: NO_x and Multi-Pollutant Control. Atlanta, EPRI TR-113187-V2, EPRI, Palo Alto, US.

Norman, J., Poukashanian, M., and Williams, A. (1997) *Fuel*, 76, 1201.

Patrick, J.W. (Ed.) (1995) Porosity in Carbons, Edward Arnold, London.

Peters, A.A.F. and Weber, R. (1993) Modelling of Swirling Natural Gas and Pulverised Coal Flames with Emphasis on Nitrogen Oxides, *IFRF Doc. No. F36/Y/21D*.

Pfefferle, L.D. and Churchill, S.W. (1989) *Ind. Eng. Chem. Res.* 28, 1004.

Pillai, K.K. (1992) Private communication.

Pitt, G.J. (1979) Coal and Modern Processing (Eds. Pitt, G.J. and Milward, G.R.), Academic Press, New York, 44.

Pourkashanian, M. and Williams, A. (1983) The Combustion of Coal-Water Slurries, *First European Conference on Coal Liquid Mixtures*, I. Chem. E. Symposium Series, No.83, 149.

Pratapas, J.M. and Holmes, J.G. (1990) Gas Cofiring in Coal-Fired Electric Generating Plants, *Proc. 7th International Pittsburgh Coal Conference*, 1051.

Raask, E. (1985) Mineral Impurities and Coal Combustion, Hemisphere Publishing

Corporation, New York.

Retcofsky, H.L. (1997) *Appl. Spectrosc.*, **31**, 116.

Rokke, N.A., Hustad, J.E., Sonju, O.K. and Williams, F.A. (1992) *Twenty-fifth International Symposium on Combustion*, The Combustion Institute, Pittsburgh, 385.

Schilling, H.D., Bonn, B. and Krauss, U. (1979) Coal Gasification, Graham and Trotman, London.

Schmid, C., Dugue, J., Horsman, H., Weber, R. (1988) *IFRF Doc. No. F259/a/5.*

Scott, D.H. and Carpenter, A.M. (1996) Advanced Power Systems and Coal Quality, IEA Coal Research, London.

Shinn, J.H. (1984) *Fuel*, **63**, 1187.

Smart, J.P., Jones, J.M., Pourkashanian, M. and Williams, A. (1997) Scale-up of Swirl-Stabilised Pulverised-Coal Burners in the Thermal Input Range 2.5–12 MW, *IFRF 12th Members Conference*, Noordwijkerhart, Netherlands.

Smart, J.P., Jones, J.M., Pourkashanian, M. and Williams, A. (1998) Scale-up of Swirl-Stabilised Pulverised-Coal Burners in the Thermal Input Range 2.5–50 MW, *IFRF Joint French/English Conference*, Guernsey.

Smith, I. (1997) Greenhouse Gas Emission Factors for Coal—the Complete Fuel Cycle. *IEACR/98*, IEA Coal Research, London.

Smith, I.W. (1982) *Nineteenth International Symposium on Combustion*, The Combustion Institute, Pittsburgh, 1045.

Smith, I., Nilsson, C. and Adams D. (1994) Greenhouse Gases—Perspectives on Coal. *IEA PER/12*, IEA Coal Research, London.

Smith, K.L., Smoot, L.D., Fletcher, T.H. and Pugmire, R.J. (1994) The Structure and Reaction Processes of Coal, Plenum Press, New York.

Smith, P., Fletcher, T.H. and Smoot, L.D. (1981) *Eighteenth International Symposium on Combustion*, The Combustion Institute, Pittsburgh, 1285.

Smoot, L.D. (Ed.) (1993) Fundamentals of Coal Combustion, Elsevier, Amsterdam.

Solomon, P.R. (1979) *Fuel ACS Div. preprints*, **24**, 184.

Solomon, P.R. and Fletcher, T.H. (1994) *Twenty-fifth International Symposium on Combustion*, The Combustion Institute, Pittsburgh, 463.

Solomon, P.R., Fletcher, T.H. and Pugmire, R. (1993) *Fuel*, **72**, 587.

Soud, H.M. and Takeshita, M. (1994) *FGD Handbook, IEACR/65*, IEA Coal Research, London.

Speight, J.G. (1993) Gas Processing. Environmental Aspects and Methods, Butterworth Heinemann, Oxford.

Speight, J.G. (1994) The Chemistry and Technology of Coal, Marcel Dekker, New York.

Spiro, C.L. and Kosky, P.G. (1982) *Fuel*, **61**, 1080.

Stach, E., Mackowsky, M. Th., Teichmuller, M. Tayler, G.H., Chandra, D. and Teichmuller, R. (1982) Coal Petrology, 3rd Edition, Gebruder Borntaeger, Berlin,

Stuttgart.

Stopford, P.J., Strachan, I.G.D. and Turrell, M.D. (1998) On-Line Boiler Performance Modelling using Computational Fluid Dynamics and Neural Networks, *AEA Technology, Control of Flames and Combustion Processes, Paper 5*, Guernsey, UK.

Stopford, P.J., Turrel, M.D. and Windsor, M.E. (1994) Numerical Simulation of an Opposed Wall Fired Power Station at Drax Fitted with Low NO_x Burners, *AEA-in-tec-1788*.

Strange, J.F. and Walker, P.L. Jr. (1976) *Carbon*, 16, 345.

Stringer, J. and Meadowcroft, D.B. (1990) *Trans. Inst. Chem. Engg.*, 68, 181.

Sugawara, A., Kurowawa, S., Hatori, H., Saito, K., Yamada, Y., Sugihara, M., Wasaka, S., Yoshida, H., Seo, T., Susuki, T., Inoguchi, M. and Sohnai, M. (1998) Combustion Conversion Technologies in the New Sunshine Program in Japan. *215th ACS National Meeting*, Division of Fuel Chemistry, American Chemical Society, 43, Dallas.

Takematsu, T. and Maude, C. (1991) Coal Gasification for IGCC Power Generation, *IEACR/37*, IEA Research, London.

Takeno, K., Tokuda, K., Ichinose, T. and Kaneko, S. (1996) *Twenty-sixth International Symposium on Combustion*, The Combustion Institute, Pittsburgh, 3223.

Thambimuthu, K. (1994) Development in Coal-Liquid Mixtures, *IEACR/69*, IEA Coal Research, London.

Tilley, H.A., O'Connor, M., Stephenson, P.L., Whitehouse, M., Richards, D.G., Hesselmann, G., Macphail, J., Lockwood, F.C., Williamson, J., Williams, A. and Pourkashanian, M. (1998) Power-Gen. Asia, New Delhi, India.

Tomeczek, J. (1994) Coal Combustion, Krieger Publishing Company, Melbourne, FL.

Truelove, J. (1984) *Twentieth International Symposium On Combustion*, The Combustion Institute, Pittsburgh, 525.

Unsworth, J.F., Barratt, D.J. and Roberts, P.T. (1991) Coal Quality and Combustion Performance, An International Perspective, *Coal Sci. and Tech.*, 19, Elsevier, Amsterdam.

Van Heek, K.H. and Mühlen, H-J. (1991) Fundamental Issues in Control of Carbon Gasification Reactivity (Eds. Lahaye, J. and Ehrburger, P.). Kluwer, Boston, p. 1.

Van Krevelen, D.W. (1993) Coal, 3rd Edition, Elsevier, Amsterdam.

Visser, B.M. and Weber, R. (1989) Computations of Near-Burner Zone Propagation of Swirling Pulverised Coal Flames, *IFRF Doc. No. F336/a/13*.

Visser, B.M. and Weber, R. (1990) Predictions of Near-Burner Zone Properties of Six Swirling Pulverised Coal Flames, *IFRF Doc. No. F036/y/14*.

Wallace, S., Bartle, K.D. and Perry, D.L. (1989) *Fuel*, 68, 1450.

Walsh, P.M., Zhang, M., Farmagan, W.F. and Beer, J.M. (1994) *Twentieth International Symposium on Combustion*, The Combustion Institute, Pittsburgh, 1401.

Weber, R. (1996) *Twenty-sixth International Symposium on Combustion*, The Combustion Institute, Pittsburgh, 3343.   238   238

Weber, R., Peters, A.A.F. and Brecthaupt, P.P. (1993) *Proc. of International Joint (ASME/EPRI) Power Generation Conference*. EPRI, Palo Alto, US.

Weeda, M., Abcouwer, H.H., Kapteijn, F., Moulijn, J.A. (1993) *Fuel Proc. Tech.*, **36**, 235.

Weeda, M., Ermers, F, v. Linden, B., Kapteijn, F. and Moulijn, J.A. (1993) *Fuel Proc. Tech.*, **36**, 243.

Wendt, J. and Pershing, D. (1977) *Comb. Sci & Tech.*, **16**, 111.

Williams, A. and Mitchell, C. (1994) Methane Emissions from Coal Mining. Mining and its Environmental Impact (Eds. Hestor, R.E. and Harrison, R.M.) Royal Society of Chemistry, Cambridge, UK.

Williams, A., Pourkashanian, M., Jones, J.M. and Rowlands, L. (1997) *J. Inst. Energy,*, **70**, 102.

Wiser, W.H. (1975) Research in Coal Technology: The University's Role (Ed. Stemberg, H.W.), Conf. Buffalo State University, New York, **II**, 57.

Wu, P.C., Lower, W.E. and Hottel, H.C. (1988) *Fuel*, **67**, 205.

Yap, L.T., Pourkashanian, M., Howard, L.M., Williams, A. and Yetter, R.A. (1998) *Twenty-seventh International Symposium on Combustion*, The Combustion Institute, Pittsburgh, 1451.

Zeng, T. and Fu, W.B. (1996) *Comb. and Flame*, **197**, 107.

APPENDIX 1
SO$_2$ and NO$_x$ Emissions Conversion Chart

To Convert	To: (Multiply by)								
From	mg/	ppm	ppm	g/GJ			lb/10⁶ Btu		
	Nm³	NO$_x$	SO$_2$	Coal[A]	Oil[B]	Gas[C]	Coal[A]	Oil[B]	Gas[C]
mg/Nm³	1	0.487	0.350	0.350	0.280	0.270	8.14 ×10⁻⁴	6.51 ×10⁻⁴	6.28 ×10⁻⁴
ppm NO$_x$	2.05	1		0.718	0.575	0.554	1.67 ×10⁻³	1.34 ×10⁻³	1.29 ×10⁻³
ppm SO$_2$	2.86		1	1.00	0.801	0.771	2.33 ×10⁻³	1.86 ×10⁻³	1.79 ×10⁻³
g/GJ Coal[A]	2.86	1.39	1.00	1			2.33 ×10⁻³		
g/GJ Oil[B]	3.57	1.74	1.25		1			2.33 ×10⁻³	
g/GJ Gas[C]	3.70	1.80	1.30			1			2.33 ×10⁻³
lb 10⁶ Coal[A]	1230	598	430	430			1		
lb 10⁶ Oil[B]	1540	748	538		430			1	
lb 10⁶ Gas[C]	590	775	557			430			1

A–Coal: Flue gas dry 6% excess O$_2$; assumes 350 Nm³/GJ.

B–Oil: Flue gas dry 3% excess O$_2$; assumes 280 Nm³/GJ.

C–Gas: Flue gas dry 3% excess O$_2$; assumes 270 Nm³/GJ. With acknowledgment to NEI International Combustion Ltd.

APPENDIX 2
Values of Equilibrium Constants

T°K	K_I $\dfrac{P_{CO}\,P_{O_2}^{1/2}}{P_{CO_2}}$	K_{II} $\dfrac{P_{H_2}\,P_{O_2}^{1/2}}{P_{H_2O}}$	K_{III} $\dfrac{P_{OH}\,P_{H_2}^{1/2}}{P_{H_2O}}$	K_{IV} $\dfrac{P_H}{P_{H_2}^{1/2}}$	K_V $\dfrac{P_O}{P_{O_2}^{1/2}}$
1000	5.69×10^{-11}	8.12×10^{-11}	5.19×10^{-12}	2.26×10^{-9}	1.66×10^{-10}
1200	1.73×10^{-8}	1.25×10^{-8}	1.62×10^{-9}	1.96×10^{-7}	2.49×10^{-8}
1400	9.74×10^{-7}	4.45×10^{-7}	9.82×10^{-8}	4.84×10^{-8}	9.40×10^{-7}
1600	1.98×10^{-5}	6.54×10^{-6}	2.16×10^{-6}	5.41×10^{-5}	1.44×10^{-5}
1700	6.82×10^{-5}	1.98×10^{-5}	7.69×10^{-6}	1.47×10^{-4}	4.44×10^{-5}
1800	2.04×10^{-4}	5.30×10^{-5}	2.39×10^{-5}	3.56×10^{-4}	1.21×10^{-4}
1900	7.69×10^{-4}	1.81×10^{-4}	6.58×10^{-5}	7.90×10^{-4}	2.96×10^{-4}
2000	1.31×10^{-3}	2.84×10^{-4}	1.64×10^{-4}	1.62×10^{-3}	6.64×10^{-4}
2100	2.91×10^{-3}	5.83×10^{-4}	3.74×10^{-4}	3.10×10^{-3}	1.38×10^{-3}
2200	5.97×10^{-3}	1.12×10^{-3}	7.90×10^{-4}	5.61×10^{-3}	2.69×10^{-3}
2300	0.01155	2.05×10^{-3}	1.57×10^{-3}	9.63×10^{-3}	4.93×10^{-3}
2400	0.0211	3.53×10^{-3}	2.94×10^{-3}	0.0158	$8.61 \text{ V } 10^{-3}$
2500	0.0366	5.85×10^{-3}	5.23×10^{-3}	0.0250	0.0144
2600	0.0610	9.33×10^{-3}	8.90×10^{-3}	0.0382	0.0231

$K_{3.1} \equiv K_I$
$K_{3.2} \equiv K_{IV}\, P_{CO_2} P_{H_2} \left(K_{III} P_{CO} P_{H_2O}\right)^{-1}$
$K_0 \equiv K_V$

APPENDIX 3
Combustion Calculations for Coal

A heat balance calculation and flue gas composition calculation are given using the following date that were obtained in a boiler trial.

Analysis of coal (dry basis):	C = 71.89%; H = 7.66%; O = 9.01%; N = 2.68%; S = 1.91%; Ash = 6.85%; Gross calorific value = 31.41 MJ/kg
Analysis of dry ashes:	Carbon 36.0%; Ash 64.0%
Average flue gas analysis:	$CO_2 + SO_2$ = 11.7%

$$O_2 = 5.5\%$$
$$CO = 1.3\%$$

Inlet air temperature	= 300 K
Exit flue gas temperature	= 578 K
Water evaporated per kg of coal	= 8.8 kg
Feed water temperature	= 348 K
Steam condition saturated at	457 K
Mean thermal capacity of flue gases	= 1.41 kL/m³ at NTP
kg molar volume	= 22.4 m³ at NTP

It is thus possible to calculate the following:

Carbon loss in ashes

Balance ash in coal to ash in "refuse":

0.64 x mass of refuse	= 6.85 kg per 100 kg of fuel
Mass of refuse	= 6.85 ÷ 0.64
	= 10.7 kg per 100 kg of fuel
Mass of carbon lost	= 10.7 × 0.36
	= 3.85 kg per 100 kg of fuel
Mass of carbon burned per 100 kg of fuel	= 68.04 kg (mass C in coal – mass carbon lost)

Combustion tabulation

The combustion equations are

$$C + O = CO_2 \quad 2H_2 + O_2 = 2H_2O \quad S + O_2 = SO_2$$

		Mol % of fuel burned	Mols oxygen required	Mols H_2O
C	$68.04 \div 12$	5.67	5.67	3.83
H	$7.66 \div 2$	3.83	1.91	
O	$9.01 \div 32$	0.28	0.28	
N	2.68			
S	$1.91 \div 32$	0.06	$\underline{0.06}$	―
			7.36	3.83

Carbon + sulfur balance

Mols carbon + sulfur per 100 kg fuel	$= 5.73$
Mols carbon + sulfur per 100 mols of dry flue gases	$= 11.7 + 1.3 = 13.0$
$0.13 \times$ mols dry flue gases per 100 kg coal	$= 5.73$
Mols dry flue gases per 100 kg coal	$= 44.1$

Carbon monoxide in flue gases

Mols carbon monoxide from 100 kg coal	$= 44.1 \times 1.013 = 0.57$

Excess air %

Mols oxygen in flue gases from 100 kg coal	$= 44.1 \times 0.055 = 2.43$
Oxygen equivalent of carbon monoxide	$= 0.57 \div 2 = 0.28$
Mols excess oxygen	$= 2.43 - 0.28 = 2.15$
% excess oxygen = % excess air	$= \dfrac{(2.15 \times 100)}{7.36} = 29.2$

Volume of wet flue gases

Total volume per 100 kg coal $= 44.1 + 3.8 = 47.9$ mols

therefore, the % contents of CO, CO_2, SO_2, and the dew point can be calculated

Heat top steam

By reference to steam tables:

Heat content of feed water at 348K	= 314 kJ/kg
Heat content of saturated steam at 457K	= 2780 kJ/kg
Total heat added in boiler per kg water	= 2780 − 314
But evaporation is 9.8 kg water per kg of fuel therefore, heat to steam	= 2466 × 8.8 kJ/kg fuel
	= 21.70 MJ/kg fuel

Additional data

By reference to tables, the following additional information can be found:

Latent heat correction of water	= 2.453 MJ/kg
Calorific value of carbon monoxide	= 12.1 MJ/m^3 at NTP
Calorfic value of carbon	= 33.7 MJ/kg

Heat balance

The overall heat balance may now be calculated per kg fuel:

Sensible heat loss	$= \dfrac{0.479 \times 22.4 \times 1.41 \times 278}{1000}$	= 4.20	13.4%
Latent heat loss	= 0.0383 × 8 × 2.453	= 1.67	5.3%
CO loss	= 0.0057 × 22.4 × 12.1	= 1.54	4.9%
Carbon loss	= 0.0385 × 33.7	= 1.30	4.1%
Heat to steam		= 21.70	69.1%
Unaccounted loss		= 0.99	3.2%
		31.40	100.0%

APPENDIX 4
Calculation of the Products of Combustion Allowing for Dissociation

The previous calculation is based on the assumption of negligible dissociation, which is the case below 1600K and is certainly the case in flue gases. Above that temperature, dissociation must be taken into account if an accurate knowledge of the flame composition is required, for example, for detailed radiation calculations.

During the combustion of pulverized coal, for example, the flame temperatures are about 2000°C (if no heat losses from the flame occur) and allowance has to be made for the dissociation of carbon dioxide and water. This can be undertaken as follows.

For 1 kmol of products at a total pressure of P (bar) we can set:

$$P_{CO} + P_{CO_2} + P_{O_2} + P_{H_2} + P_{H_2O} + P_N = P$$

$$K_1 = \left(P_{CO}\, P_{O_2}^{1/2}\right) / \left(P_{CO_2}\right)$$

$$K_2 = \left(P_H\, P_{O_2}\right)^{1/2} / \left(P_{H_2O}\right)$$

$$k = n_C/n_O$$

$$l = n_H/n_O$$

$$m = n_N/n_O$$

where n_O, etc., are the number of k atoms of oxygen in 1 km of combustion gases:

$$Pkn_O = P_{CO} + P_{CO_2}$$

$$Pln_O = 2P_{H_2} + 2P_{H_2O}$$

$$Pmn_O = 2P_{N_2}$$

and therefore,

$$Pn_O = P_{CO} + 2P_{CO_2} + 2P_{O_2} + P_{H_2O}$$

$$n_O = \frac{\left(1 - p_{O_2}/P\right)}{k + l/2 + m/2} = \frac{\left(1 - p_{O_2}/P\right)}{z}$$

where $z = k + l/2 + m/2$.

Setting $c = p_{O_2}$ in order to write the following equations more conveniently, the iteration used is

$$c_{n+1} = C_n^2/P$$

$$\left[\frac{k}{(K_1/c^{1/2}) + 1} + \frac{l/2}{(K_2/c^{1/2}) + 1} \right] + k - 1 = \left[\frac{\left(1 - c_n/P\right)^2}{2z} \right]$$

The partial pressure of oxygen (c_n) has to be guessed (almost any value can be used, but the better, the faster the iteration) and the iteration undertaken by a simple computer program (or by hand), and it usually converges within 10 iterations.

After iteration the remaining partial pressures are calculated from the following equations:

$$n_O = (1 - c/P)z$$

$$P_{CO} = Pkn_O K_1/(K_1 + c^{1/2})$$

$$P_{CO_2} = P_{CO} c^{1/2}/K_1 \quad \text{or} \quad Pkn_O - p_{CO}$$

$$P_{H_2} = Pln_O K_2/(2K_2 + c^{1/2})$$

$$P_{H_2O} = P_{H_2} c^{1/2}/K_2$$

$$P_{N_2} = Pmn_O/2$$

This calculation is sufficient for most practical calculations. If more extensive dissociation takes place, as occurs in flame fronts or in high-temperature (oxygen enriched) flames, then high concentrations of radicals result, requiring a full computation. This can be undertaken by algebraic methods or by minimization of the Gibbs free energy (Kuo, 1996).

A very considerable number of computer programs are available to undertake this type of calculation including CHEMKIN (Sandia Laboratories), FLAME (AEA Technology), and EQUITHERM (VCH Publishers). These can also be used for flame temperature calculations (see Appendix 5).

APPENDIX 5
Flame Temperature Calculations

Adiabatic flame temperature can be calculated for the dissociated or undissociated flame case as follows:

$$T(\text{K}) = \frac{Q_n + P_r - \Delta H_d}{.m_p \, \overline{C_{p(p)}}} + 298$$

where Q_n is the net heat of combustion of the fuel (at 25°C), $.mp$ is the mass of products per unit mass of fuel, and $C_{p(p)}$ is their specific heat. P is the degree of air preheat above 25°C and is given by $m_a(C_{p(a)}T_a)$, where m_a is the mass of air, and $C_{p(a)}$ is the mean specific heat of air (units of mass). ΔH_d represents the heat adsorbed by the dissociation reactions and is zero in the case of the commonly used undissociated flame temperature calculation. In order to calculate ΔH_d, the equilibrium composition of the products at the flame temperature is required (see Appendix 4).

The above equation can be solved using this information together with mean specific heat data over the range of temperatures. The flame temperature can be obtained more accurately using specific heats or enthalpies over the temperature range by an iterative method. C_p data can be obtained from, e.g., JANAF tables. Solution of the flame temperature is then followed using the procedure in Kuo (1986). A number of computer programs are available to undertake this operation as outlined in Appendix 4.

NOMENCLATURE

A_0, A_1, A_2	empirical constants of devolatilization
A_f	frequency factor
A_g	specific pore surface area of the char
A_t	reactive surface area of the char
C	carbon content (daf) basis
C_i	specific heat of species i
C_o	active oxygen concentration
d	diameter
D_0	binary diffusion of oxygen in nitrogen
D_p	diameter of particle
E, E_0, E_1, E_2	activation energy
E_a	apparent activation energy (char combustion)
E_{actual}	actual activation energy (char combustion)
E_{air}	excess air
EI_{NO_x}	emission index
f_v	current volatile mass fraction
f_{mac}	maceral correction factor
F	firing rate
F_a	surface tension
F_g	gravitational force
F_r	rotational force
g	gravitational acceleration
H	hydrogen content of char
H_c	enthalpy of combustion
H_i	enthalpy of species i
H_{char}	hydrogen content of char
HHV	higher heating value
I	combustion intensity
In_{LR}	fraction of high reflectance (less reactive) vitrinite
In_R	fraction of low reflectance (reactive) vitrinite

k_i	reaction rate coefficient for species i
LLV	lower heating value
m_{ash}	mass of ash
m_i	mass of species i
m_p	particle mass
m_{p0}	initial mass of particle
m_{v0}	initial mass of volatiles
M_i	molecular weight of species i
P	pressure
q	equivalence ratio
Q	air quality
r_{ash}	radius of ash layer on char particle
R	universal gas constant
R_1, R_2	devolatilization rate
R_{ash}	ash film diffusion constant; propensity of ash to shed from char particle
R_c	chemical reaction rate coefficient per unit external surface area
R_{diff}	diffusional reaction rate coefficient
R_p	radius of unreacted core particle
R_T	overall reaction rate
R_x	rate of release of species x
t	time
T	temperature
T_0	initial temperature
T_p, T_g	temperature of particle and gas, respectively
U_0	initial velocity
V^*	total fraction of volatiles in the particle
V	fraction of particle already released as volatiles
V_c	combustion volume
Vit_M	fraction of matrix vitrinite
Vit_{PS}	fraction of pseudo vitrinite
w	weight
X_i	mole fraction of species i
α, β	constants
α_1, α_2	user-supplied weighting factors
γ	ash-particle surface tension
ε	porosity of the ash layer

η   reaction order
ρ   density
σ_t turbulent Schmidt number
σ_a apparent density of the char
φ   ratio of reacting surface with external (equivalent sphere)
                    surface area of particle
φ   mechanism factor (equals 2 at high temperatures, 1 at low
                    temperatures)
Φ   fluctuation of a scalar quantity
ω   angular velocity
θ   contact angle of ash with the char surface
γ_P characteristic size of particle
μ_t turbulent viscosity

CONVERSION FACTORS

1 tonne of oil equivalent, TOE (net, low heat value)	= 42	GJ
1 tonne of coal equivalent, TCE (standard, LHV)	= 29.3	GJ
1000 m^3 of natural gas (standard, LHV)	= 36	GJ
1 tonne of natural gas liquids	= 46	GJ
1 TCE	= 0.697	TOE
1000 m^3 of natural gas	= 0.857	TOE
1 tonne of natural gas liquids	= 1.096	TOE
1 tonne of fuel-wood	= 0.380	TOE
1 tonne of uranium (current-type reactors, open cycle)	= 8,000	TOE
1 tonne of uranium (breeders)	= 500,000	TOE

1 barrel of oil = approx. 0.136 tonne
1 cubic foot = 0.0283 cubic meter

1 calorie (cal)	= 4.196	J
1 Joule (J)	= 0.239	cal
1 Btu	= 252	cal
1 Btu/lb	= 2.326	kJ/kg

INDEX

A

Abrasion index, 38
Acceptor regeneration, 136–37
Acetylenes, 58
Acid rain, 5–6, 10
Acid smuts, formation of, 69–71
Acidic gases, 9–10
Adiabatic flame temperatures, 156, 253
Advanced clean coal technologies, 15–19
Afterburning, 66
Air gasification, 211
Allochthonous swamps, 25
Aluminum, 62
Ammonia, 237
Anthracite briquetting—Ancit process, 209
Anthracite coals, 6, 41
Anthracite-mild heat treatment process,
    209–10
Ash, 10, 30–31, 236
    fusibility of, 37
    and slag deposition, 158–60
Atmospheric fluidized-bed combustion,
    130–40
Atmospheric fluidized-bed modeling,
    133–35
Atmospheric pollution, 6
Atmospheric pressure fluidized-bed combus-
    tion (FBC) boilers, 160–64
Atomization of coal-water slurries, 197–99
Australian classification, 49

B

Babcock Hitachi Process, 71
Badzioch and Hawksley model, 100
Basic gasification reactions, 215–18
Bed composition, 137
Bed depth for fluidized-bed boilers, 163–64
Biomass or waste, co-firing of coal with,
    187–91
Bituminous coals, 6, 12, 40–41
    chemical composition of, 12
Boudouard reaction, reaction mechanism
    and intrinsic kinetics of, 233–35
Bright coals, 41–42
Briquettes and smokeless fuels, formation
    of, and their use, 208–10
British Coal (NCB) Classification, 47
British Coal (NCB) Ranking method, 47
Brown coal, 40

composition of, 12
Brown coals, 40
Bubbling fluidized-bed combustion (BFBC),
    130

C

Calcium, 62
Calcium sulfate, 137
Calcium/sulfur ratio, 163
Calorific value, 35
^{13}C and ^{1}H nmr studies, 52
Carbon, 33, 57
    formation of unburned, 60
Carbon burn-out, 144
    development of carbon burn-out models
        for high levels of, 110–22
Carbon dioxide (CO_2), 9, 10, 66
    reduction in emissions, 19
Carbon-in-ash, 58
Carbonized wood, 44
Carbon loss in ashes, 248
Carbon monoxide (CO), 10
    in flue gases, 249
    formation of, 65–66
CBX model, 181
Cellulose, 25–26
Cenosphere formation, 60
Chain burning, 126–28
Char burn-out, 88, 101–10
Char reaction, nitrogen oxides (NO_x) from,
    118–19
Chemical cleaning, 195
Chemiluminescent methods, 155–56
CHEMKIN, 252
Chlorinated fluorocarbons (CFCs), 10
Circulating fluidized-bed-combustion tech-
    nology (CFBC), 131, 133
Circulating fluidized-bed systems, 164–65
Clarain, 42, 43, 44
Clarite, 43
Coal
    aromatic nature of, 2
    blending of, and coal cleaning, 144
    characterization of, 28–39
    chemical structure, 52–54
    classification, 39–51
    co-firing of, with biomass or waste,
        187–91
    combustion calculations for, 248–50

Coal (*continued*)
  composition of, 1–2
  conversion of wood to, 25–28
  formation, 21–28
  mineral matter in, 44
  moisture in, 29–30
  origin of, 1–2
  size, 143–44
  as source of energy, 2–11
  storage, 143–44
  vegetable origin of, 21, 23
Coal ash, formation of, 62–65
Coal-based combined cycle technologies, 18
Coal benefication, 14
Coal blending, 143–44
Coal cleaning and blending of coals, 144
Coal codification, 49
Coal combustion
  industrial applications of, 142–86
  modeling, 87–88
  two-component, 187–210
  types of, 11
Coal devolatilization, 88
  computational models for, 94–99
Coal-fired boilers (PCF), 147
Coal gasification, 211–15
Coalification, 1
  conversion of wood to coal, 25–28
Coal industry, 6
Coalite process, 209
Coal-oil mixtures (COM), 193, 207
  combustion of, 207–8
  production of, 207
Coal particles
  combustion of, 59–60
    in a fixed, moving, or fluidized bed,
      123–41
  devolatilization of, 89–100
Coal petrology, 41–46
Coal production, 6
Coal pyrolysis products, 91–92
Coal seams
  formation of, 23, 25
  manner of occurrence, 23
Coal-swelling tests, 34–35
Coal utilization, 11, 12, 56
Coal-water fuel (CWF), 193
Coal-water mixtures (CWM), 193
Coal-water slurries, 193
  atomization of, 197–99
  combustion of, 192–208
    fluidized-bed, 206
    mechanism, 199–204
    theoretical modeling of, 204–6
  development of, 192–97
Coal-water-slurry spray flames, 206

Co-firing
  of coal with biomass or waste, 187–91
  with natural gas, 192
Combustible sulfur, 34
Combustion, 170–72
  calculation of products of, allowing for
    dissociation, 251–52
  general nature of coal, 11–14
Combustion calculations for coal, 248–50
Combustion chamber performance, 158
Combustion intensities, 157–58
Combustion plant, modeling of, 179–82
Combustion process, methods of reducing
    emissions of particulate materials
    by controlling, 60–62
Combustion stoichiometries and flame tem-
    peratures, 150, 154–57
Computational codes, development of, 88
Computational fluid dynamics methods,
    170–78
Constant-residence-time scaling, 167
Constant-velocity scaling, 167
Conversion factors, 257
Corrosion deposits, 159
CPD, 96, 97

D
DEMKOLEC, 19
Desmoke then briquette (housefire process),
    209
Destec entrained-flow gasifier, 225
Detailed reaction mechanisms and intrinsic
    kinetics, 228, 230–35
Devolatilization, 12, 58, 60, 85
  of coal particles, 89–100
Direct numerical simulations (DNS), 173
Discrete Transfer Model, 172
Durain, 42, 43

E
Economic boiler, 145–47
Economizer, 146–47
Eddy breakup models (EBU), 177
Electrostatic precipitator (ESP), 65
Elementary analysis, 33
Elutriation, 131
Emissions, computation of, 178–79
Energy
  coal as source of, 2–11
  from waste, 187
Engineering calculation method, 173
Entrained-flow gasification reaction,
    222–25
Entrained-flow gasifiers, 224
Environmental impact, 9
Environmental problems, 5–6

Equilibrium constants, values of, 247
EQUITHERM, 252
Equivalence ratio, 154

F

Favre averaged equations, 174–75
Field ionization mass spectrometry (FIMS), 92
Fixed- and moving-bed combustion, 123–30
Fixed-bed gasification processes, 220–22
Fixed carbon, 32
Fixed or traveling beds, combustion on, 145–47
FLAME, 252
Flame temperatures, 156–57
    calculations of, 253
FLASHCHAIN, 96–97, 181
Flue gases, 14
    carbon monoxide in, 249
    composition, 155–56
    treatment, 80–81
Flue stack losses, 157
Fluidized-bed combustion, 14, 160–66
    of coal-water slurries, 206
    commercial development of, 137–38
Fluidized-bed processes, 225, 228
Fluidized beds and stokers, 185–86
Fluidized-bed systems, 17–18
Fly ash, 62–63
Fourier transform infrared (FTIR) spectroscopy, 52
Fourier transform infrared (FTIR) technology, 37
Fuel-nitrogen (fuel-N) compounds, 76–78
Fuel-nitrogen release, 116–17
Fuel-NO, 115–16
Functional group-depolymerization vaporization cross-linking (FG-DVC) computational model, 96–97, 100, 181
Fusain, 42, 43

G

Gas cleanup, 236–37
Gasification, 55, 191
    basic reactions, 215–18
    general nature of coal, 11–14
    methods, 219–28
Gas turbine requirements, 237
Grate and fluidized-bed combustion, 190
Gray-King coke test, 34–35
Greenhouse gases, 9, 10–11
    emissions, 19–20
Gross calorific value (GCV), 35

H

Hardgrove index, 38
Heat balance, 250
Heat top steam, 250
Highly decayed "mud," 44
High-pressure particulate removal (HGCU), 139
High-temperature fouling deposits, 159
High Temperature Winkler technology, 19
Hydrogen, 33

I

Industrial applications of coal combustion, 142–86
Industrial burners, general features of, 149–50
Infrared spectra, 37
Inherent moisture, 30
Integrated gasification combined cycle (IGCC) technology, 212
International Coal Classification Scheme (ISO), 47, 49–51
Iron, 62
IR spectroscopy, 52
Isocyanic acid (RAPRENO$_x$) process, 82

J

Japanese air-blown entrained gasifier, 225

K

Kjeldahl method, 33
KOBRA project, 19
Koppers-Totzek (KT) process, 223

L

Lancashire boiler, 145
Large eddy simulation (LES), 173
Leading-order scaling law, 169
Lignins, 25–26
Lignites, 40
    composition of, 12
Loss on ignition (LOI), 58, 154–55
Lower calorific value (LCV), 35
Low-nitrogen oxides (NO$_x$) burners, 150
Low-price emission control technologies, 149
Low-temperature fouling deposits, 159
Low-temperature slag deposits, 159
Lurgi gasifier, 19

M

Macerals, 43
Magnesium, 62
Methane, 9, 10
Microlithotypes, 43
Mineral matter, 44

in coal, 44
Moisture in coal, 29–30
Molecular nitrogen, 87
Monte Carlo method, 172

N

Natural gas, 8–9
  co-firing with, 192
Navier-Stokes equations, 173
Nitrogen, 33
Nitrogen oxides (NO$_x$), 5, 9–10, 14, 182
  from char reaction, 118–19
  emissions
    reducing, 14
    scaling of, 167–70
  emissions conversion chart, 246
  formation
    control of, by combustion modifica-
      tion, 78–80
    interaction of sulfur oxides (SO$_x$) on,
      82–83, 121
  formation and control of oxides of,
    73–85
  global emissions of, 86
  mechanism of formation of, 73
  pollutant formation, 112
  scaling, 168
Nitrogen oxides (NO$_x$) models, developments to, 117–19
Nomenclature, 254–56
Noncaking coals, 204
North American Coal Field, 45–46
Nucleation, 58

O

Oil burners, 150
Organic sulfur, 34, 120
Overfeed, 123
Overfeed burning, 124, 126
Oxygen-based integrated gasification com-
  bined cycle (IGCC) technology,
  18–19
Oxygen-gasification, 211, 219

P

Partially incomplete combustion products
  (PIC), 10
Particle sizing, 35
Particulate materials
  formation and control of, 57–65
  methods of reducing emissions of, by
    controlling combustion process,
    60–62
Pf, equipment for combustion of, 149–52
Pf boilers, 147
Pf burners, 150

Phosphorus, 25
Photochemical smog, 10
Phurnacite process, 209
Plant proteins, 25–26
Pollutants, 10
  emissions of, 57
Polyaromatic hydrocarbons (PAH), 59
Polyethynes, 58
Power plants, equipment for combustion of
  pf in, 149–52
Power station and other boilers, 182–85
Prenflo process, 225
Pressurized circulating fluidized-bed com-
  bustion (PCFB), 140–41
Pressurized fluidized-bed combustion
  (PFBC), 138–40, 166
Pressurized pulverized coal combustion
  (PPCC), 16–17
Presumed pdf shape, 177
Process problems and environmental con-
  siderations, 235–37
Prompt-NO, 114
Prompt-NO route, 75–76
Proximate analysis, 29–32
Pulverized coal, 142, 143–44
  combustion of, 15–16, 86–122
Pulverized coal-fired (PCF) boilers, 147–48
Pulverized-fuel combustion, 190–91
  role of, 86–88
Pyretic sulfur, 119, 120
Pyrite, 144
Pyritic sulfur, 34
Pyrolytic decomposition, 12

R

Radiative heat transfer, 171–72
Rank, 1
  classification by, 44–46
Reaction mechanism and intrinsic kinetics
  of Boudouard reaction, 233–35
Reaction mechanism and intrinsic reaction
  kinetics of water-gas reaction,
  230–33
Reactivity, 105–6
Reburn
  modeling, 122
  as nitrogen oxides (NO$_x$) control strat-
    egy, 83–84
Reflectance, 37

S

SANDIA OPP-DIFF computer programs,
  101
Scaling criteria for burners and furnaces,
  166–70
Semibituminous coals, 41

Seyler diagram, 45–46
Shell boiler, types of, 145–47
Shell gasification process, 224–25
Silica, 62
Slag deposits, 159
Slagging, 144
Slurry preparation, 194–97
Smoke, 57
Smoke formation, features of, 57–60
Sodium, 62
Solvent-refined coals (SRC), 193
Soot, 57, 58–59
Soot-forming conditions, 59–60
Split flame burner, 150, 152
Stack solids, 57
Steam boiler applications, 150, 152
Subbituminous coals, 28, 40
Substitute natural gas (SNG) production, 218
Sulfur, 25, 34
Sulfur dioxide (SO_2), 5, 14
    control of emissions, 71–73
    formation of, 67
Sulfur emissions, 119–21
Sulfur oxides (SO_x), 9, 14
    emissions conversion chart, 246
    interaction of, on nitrogen oxides (NO_x) formation, 82–83, 121
Sulfur present in coal, pollutants originating from, 67–73
Sulfur retention, 135–36
Sulfur trioxide, formation of, 67–69

T
Tar, 236
Texaco process, 225
Thermal decomposition, 137
Thermal $deNO_x$ selective noncatalytic reduction (SNCR), 81
Thermal gravimetry, 35, 37
Thermal-NO, 112–13
Thermal power stations, combustion chamber in, 150, 152
Thermal route, 73–75
Total moisture, 30
Traveling-grate burning, 126–28

Traveling-grate combustion, 123
Turbulence-combustion interaction, models for, 176
Turbulent combustion, governing equations of, 174–76
Turbulent combustion modeling, 172–73
Two-component coal combustion, 187–210

U
UK National Coal Board (NCB) classification, 46–47, 48
Ultimate analysis, 33
Unburned hydrocarbons (UHC), 10
Underfeed, 123
Underfeed burning, 128–29
Urea, 82
US (ASTM) coal classification system, 47, 49

V
Vegetable origin of coal, 21, 23
Vitrain, 42, 43
Vitrite, 43
Volatile matter, 31–32
    and caking power, 33–34
Volatile organic compounds (VOC), 10
Volatile-released NO, 117–18
Volatile release-rate representation, 100
Volatiles, combustion of, 101

W
Water-gas reaction, reaction mechanism and intrinsic reaction kinetics of, 230–33
Water-tube boilers, 147
Wet flue gases, volume of, 249
Wood, conversion of, to coal, 25–28
Woody tissue, 44

X
XANES, 52
XPS, 52

Z
Zeldovich mechanism, 73
Zeroth-order scaling, 169